City and Cosmos

City and Cosmos

The Medieval World in Urban Form

KEITH D. LILLEY

REAKTION BOOKS

For Mum

Published by Reaktion Books Ltd
33 Great Sutton Street
London EC1V 0DX, UK
www.reaktionbooks.co.uk

First published 2009

Printed and bound in China

British Library Cataloguing in Publication Data

Lilley, Keith D.
City and cosmos: the medieval world in urban form
1. Cities and towns, Medieval
I. Title
307.7'6'094'0902

ISBN: 978 1 86189 441 0

CONTENTS

INTRODUCTION:
THE CITY-COSMOS IDEAL
7

PART I – CITY-COSMOS IMAGINED
1 URBAN MAPPINGS
15
2 URBAN FORMS
41

PART II – CITY-COSMOS BUILT
3 FOUNDING A CITY, FOUNDING A WORLD
77
4 MEASURES OF MEANING
95

PART III – CITY-COSMOS LIVED
5 MORAL TOPOGRAPHIES
131
6 PERFORMING BODIES
158

EPILOGUE:
CITIES OF GOD?
185

REFERENCES 189
SELECT BIBLIOGRAPHY 231
ACKNOWLEDGEMENTS 246
PHOTO ACKNOWLEDGEMENTS 247
INDEX 248

INTRODUCTION

The City-Cosmos Ideal

What shall we say of the city?
ST AUGUSTINE OF HIPPO, *The City of God* [1]

In describing his various 'categories of medieval culture', Aron Gurevich notes that 'the analogy between microcosm and macrocosm lies at the very root of medieval symbolism.'[2] It is an analogy which forms the foundation of this book too, for it is what gave the medieval city its symbolic meaning. In the Middle Ages everything in the Christian world had its rightful place in a divinely ordained and ordered hierarchy that extended through from the cosmic 'body' above, the macrocosm, to the human body below, the microcosm. The city too was understood to be part of this hierarchy. It came to be seen both as a microcosm of the universe and a macrocosm of 'man', an urban 'body' placed between the macro and microcosmoi. These microcosmicisms were closely entwined in Christian thinking of the Middle Ages and both were put to use conceiving the city as a world *in parvo* and as a body writ large.[3] In this sense, the city was simultaneously an image of the universe and of man, both having been created and ordered by God himself. Moreover, these principles are to be found throughout the Middle Ages in how cities and urban spaces were *imagined*, *built* and *lived*.

Plato's *Timaeus* is the foundation of this medieval cosmology and microcosmicism.[4] The *Timaeus* provided the secure cosmological underpinnings on which Plato had sought to construct his ideal city-state. His parallels between the ordering of the 'cosmic body' and the ordering of the 'human body' were to be made time and again in Neoplatonic cosmologies of medieval Europe, and so came to occupy the minds of later Christian thinkers throughout the Latin West.

Aristotle made use of microcosmicisms, too, and they likewise circulated around the medieval Christian world through the many translations and commentaries that were made of cosmological texts such as the *Physics* (the *Timaeus* circulating throughout the early and high Middle Ages, the *Physics* rather later, thanks largely to Aquinas' commentary of the mid-thirteenth century). Through such Platonic and Aristotelian works on natural philosophy, as well as political philosophy, such as Aristotle's *Politics* and its derivatives, threads of Classical microcosmicism were woven into the fabric of medieval Christian thought.[5]

A translation of and commentary on Plato's *Timaeus* by Calcidius 'provided the Middle Ages with a direct contact with Plato's enigmatic but fascinating work', particularly during the eleventh and twelfth centuries.[6] As Klibansky notes, 'there was hardly a medieval library of any standing which had not a copy of Calcidius' version' of the *Timaeus*.[7] It was through Calcidius particularly, then, that Neoplatonic microcosmicism became familiar in the Latin West (illus. 1). This prominence of Neoplatonism in Christian thinking in the high Middle Ages has of course attracted enormous debate in recent years, with particular dissent over the authorship of individual manuscript glosses and commentaries on the *Timaeus* and the question of the influence of the 'school' of Chartres and whether there was anything definably 'Chartrian' in its genre of scholasticism.[8]

For Calcidius, not only is the city a small world – an ordered microcosm of the universe – it is also a body writ large – a macrocosm of the human being; all ordered and functioning according to the same principles and overall design, which in the city are present in its spatial form as well as in its social order.[9] Calcidius thus explained:

> In the nature of man there is something royal, another is placed in the middle, and a third is set in the lowest part; it is the highest which commands, the middle which acts, and the third which is governed and ruled. Therefore the soul commands, the energy placed in its breast carries out, and the other parts of the body down to the genitals and below are ruled and managed.[10]

By analogy, and in a metaphorical and mystical sense, this tripartite scheme is seen both to order the cosmos and the human body, as well as the social world of the imagined city-state:

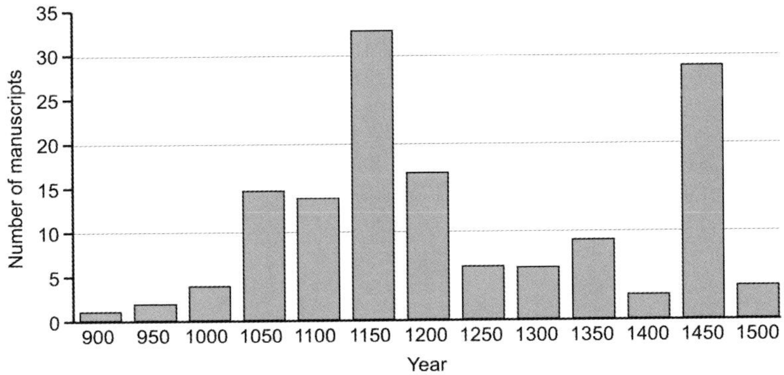

1 Numbers of surviving manuscripts of Calcidius' *Timaeus* over fifty-year periods, 900–1500.

> He ordered the chief men of this city as the most prudent and wise to dwell in the highest places of the city, under these were to live the young soldiers outfitted with arms, to whom he subjected the tradesmen and masses, so that the chief men as ones full of wisdom might give orders, the soldiers might act and carry out, and the masses might furnish appropriate and useful service. So we see the soul ordered; indeed its rational part as the wisest holds the chief part just like the Capitol of the whole body [i.e., the head], its energy which is similar to temper is like young soldiers abiding in camps of the heart, and the masses and tradesmen, that is desire or passion, are hidden in the lower regions and are concealed by nature.[11]

With this schema, then, Calcidius connects cosmos-city-body, a microcosmicism that, as Dutton demonstrates, had an important influence on Christian Neoplatonists later in the Latin West, particularly in the eleventh and twelfth centuries.[12]

The continued influence of the Timean microcosmic model in the Latin West can be judged, for example, by later glosses and commentaries by those who in their own work showed an interest in its cosmographical outlook. With Alan of Lille the evidence for this reformulation of the Platonic 'city-cosmos' model comes not only through his *Plaint of Nature*, but also, interestingly, through a series of surviving sermons inspired, it seems, by Calcidius' *Commentary*. In the *Plaint* comes the following observation:

> See how in this universe, as in a noble state, a certain excellence of administration is established by an approved plan of

> management. For in heaven, as at the pinnacle of an earthly state, the eternal commander has his imperial abode; from all eternity the order had gone out from him that each and every thing should be inscribed and made known in the book of his Providence. In air, as in the centre of the city, the celestial army of angels carry on the battle and, in the capacity of deputies, assiduously extend their protection to man. Man, however, like a foreigner, living on the outskirts of the universe, does not refuse to show obedience to the hosts of angels. In this state, then, God gives commands, the angels carry them out, and man obeys.[13]

The emphasis here, then, is on the cosmos – the wider universe – being presented as a 'city', or 'state', the one having the form and order of the other. Following on from this passage, Alan turns to show that man too, as a microcosm himself, is also a 'city':

> The image of this perfectly organized state shines forth in man. Wisdom, that gives orders, rests in the citadel of his head and the other powers, like demi-goddesses, obey her as if she were a goddess. For native ability, power of reasoning, as well as the faculty of recalling the past dwell in the various compartments of the head and obey Wisdom with enthusiasm. In the heart, as in the middle of the earthly city, Magnanimity has taken up her abode; she has sworn military allegiance to Wisdom, as commander-in-chief, and carries out operations according to the decisions of Wisdom's command. The loins like the city's outskirts, give the lower portions of the body wilful desires; they do not dare to oppose the orders of Magnanimity but obey her will. In this state, then, the role of commander is assumed by wisdom, the likeness of administrator by Magnanimity, while desire appropriates the image of the one obeying. In other things, too, the form of the human body takes over the image of the universe.[14]

Alan's 'city' is thus both cosmos and body, each formed and functioning according to divine plan. His city has the same *moral topography* as the universe, with its highest orders at the centre and lowest at the edge, and as the human body. Similarly, William of Conches also made use of the tripartite scheme in his gloss on the *Timaeus*, as follows:

> According to this likeness, Socrates wanted there to be a senate in the citadel of the city as wisdom is in the citadel of the head, under it soldiers as courage abides in the heart, and under the soldiers confectioners as desire is in the loins. And as heavy feet in the lowest part of the body tread the earth, so farmers, hunters and shepherds manage the land outside the walls of the city.[15]

William's rendering of the *Timaeus* is clearly close to Calcidius' own words. He elaborates further on this theme in his comments in his gloss on Macrobius' *Commentary on the Dream of Scipio*, another popular Neoplatonic cosmography:

> In the state Plato wanted there to be some men commanding and overseeing such as the senate, some fighting such as soldiers, and some serving such as the lower classes. Plato and Socrates, in their arrangement of the state, imitated the divine arrangement which in the human body is the following: the head holds the highest place and, so to speak, lordship over the other members, and in it is the abode of wisdom on account of its three chambers, which we will discuss elsewhere. Therefore, just as wisdom is located in the head and oversees the remaining members of the body, so the senators being in the highest place, that is, in the citadel of the city, oversee the lower classes and regulate their movements and actions. Under the head are the hands which are disposed to act and the heart where the abode of courage is located; so under men from the senatorial order are soldiers disposed to endure hardship and are courageous in defence of the state. The kidneys, in which human desire flourishes, are located under the heart and so under soldiers are found confectioners, cobblers, skinners, and other craftsmen. At the remotest part of the body, feet are found, so outside the walls on the outskirts of the city are farmers to cultivate the fields.[16]

The reformulation of the Timean city-cosmos model by Alan of Lille and William of Conches thus encompasses not only those who were within the 'walls of the city' but also those without – but it is the city that is positioned closer to God, being of a higher order; those in 'the citadel of the city' being 'highest' of all. This 'divine arrangement' – a model of celestial geography – is imprinted upon the very layout of the city through its moral topography. Cosmic hierarchy therefore takes on a specifically urban form, spatialized through the arrangement of constituent parts just as corresponding 'members' of the human body are.

In the discussions of these Latin Christian thinkers, then, of which there were many, the city was not simply seen as a set of buildings and people, but rather a 'map' of Christian beliefs about the wider world, about cosmology and cosmogony. Despite an awareness of twelfth-century Neoplatonists and their discussions of Calcidius' *Timaeus*, these ideas about the city and its cosmological symbolism have so far been overlooked by successive historians of medieval urbanism.[17] Yet such ideas were circulating throughout the Middle Ages in Europe, connecting local and global, earthly and spiritual worlds, and universal and particular, material and imagined cities. Through them the city itself was understood as a microcosm of the wider world as well as a macrocosm of the human form.

Thinking of the 'city' both as a macro- and microcosm gave urban spaces meaning in the medieval imagination. The purpose of this book and its following chapters is to explore this further, tracing how these themes and traditions were reflected and reinforced through imagining, building and inhabiting towns and cities in medieval Europe. It offers a new perspective on medieval urbanism, attempting to understand the city as a cultural as well as a material construct, and to see it more from the point of view of those who knew it and experienced it first-hand. What is presented here is another reading – or gloss – on urban life in the Middle Ages, therefore; not so as to supplant existing and accepted interpretations but rather to supplement and enhance them. The medieval city, it will be seen, acquired its cosmological symbolism both through its spatial *forms*, for example in the ordered geometries it shared with the cosmos, and in its *functions*, as a 'body' made up of hierarchically arranged parts mirroring the moral topography of the Christian universe as a whole, all created to God's divine plan.

PART I

City-Cosmos Imagined

ONE

URBAN MAPPINGS

FROM THE HEAVENLY TO THE EARTHLY CITY

The city that lay at the heart of the medieval world – the *axis mundi* – was of course Jerusalem. 'The city of Jerusalem I have set among the nations', Ezekiel proclaims, 'with the other countries round about her.'[1] Just as the round earth was at the centre of a circular Christian cosmos, so the city of Jerusalem was the symbolic centre of the world.[2] The Psalter map, for example, shows Jerusalem as a round red spot encompassed by two concentric rings (illus. 36).[3] This particular form is also used in more detailed visual representations of the city, which typically depict it as a circle of walls. This idealized form of Jerusalem was, in Ousterhout's words, a 'flexible geography and transportable topography',[4] in which the holy city was presented as an archetype, an imagined vision of what the city looked like, rather than how it actually was. To this end, the circular-shaped Jerusalem not only symbolized a wider cosmos, of which it was the sacred and spiritual centre, but it also provided a model on which to fix images of other cities. Jerusalem's idealized geometrical form became the basis for depicting cities all over the medieval Christian world.

Heavenly Jerusalem, the celestial city, is described in Revelation descending down from heaven: 'and I John saw the holy city, New Jerusalem, coming down from God out of heaven, prepared as a bride adorned for a husband.'[5] The city's richness is conveyed in colourful, beautifully drawn images that illustrate the many medieval manuscripts relating the story of the Apocalypse, an imagery that emerged in the fourth century and gained in popularity through the Middle Ages, particularly the Carolingian period, but also later on in the thirteenth century.[6] In all these images the Heavenly Jerusalem is depicted slightly differently in its details. Sometimes it is populated with buildings and turrets, as with examples in the Trier and Cambrai manuscripts of the ninth century (illus. 37); sometimes it is simply a space enclosed by walls devoid of detail except for Christ, signified usually either as the Lamb of God or the Tree of Life positioned at its centre, as shown in the Trier, Cambrai and Valenciennes manuscripts (illus. 38).[7] Sometimes the city is shown in elevation, as a bird's-eye view, and sometimes projected as a plan, 'laid flat'.[8]

The Biblical description of the heavenly Jerusalem makes clear what shape the celestial city should be: 'the city lieth four-square, and the length is as large as the breadth', with 'on the east three gates; on the north three gates, on the south three gates; and on the west three gates.'[9] Such a square-shaped city is depicted in the thirteenth-century Trinity Apocalypse manuscript, and the earlier Paris Bibliothèque Nationale manuscript 2290, but in Carolingian drawings of the ninth century the heavenly city is invariably presented as circular in form, a circle of walls, as indeed it is in some later sources, such as the 'Trinity Apocalypse' and Lambert of Saint-Omer's *Liber floridus* (illus. 39 and 40).[10] Depictions of a circular-shaped celestial Jerusalem are clearly at odds with the city's square form described by scripture. The reason for using a circle seems to lie in the cosmological significance of its geometrical form, coupled with the central place of Jerusalem in the Christian world.[11]

There was an imagined symmetry between the shape of Jerusalem and the shape of the world, as Ousterhout has noted: 'the same schema that underlies the medieval world maps defines the plan of Jerusalem – an "O" forming the walls of the city is divided by a cross or a "T" of the main streets', for in each image 'God's order is expressed in geometric terms' (illus. 36).[12] The circular-shaped city was itself the innermost of a series of hierarchically arranged concentric rings that organized the whole of the Christian world spatially and geometrically from its centre, Jerusalem, outwards to the very edge of the celestial sphere. Hence the Jerusalem located at the centre of the Hereford world map (*c.* 1290) has the same topographical details as those found on Crusader maps of the city from the twelfth century (illus. 2).[13]

So although examples such as the Trinity Apocalypse manuscript show a square plan of the heavenly Jerusalem, as it is described in Scripture, an idealized form was also used depicting the city as a

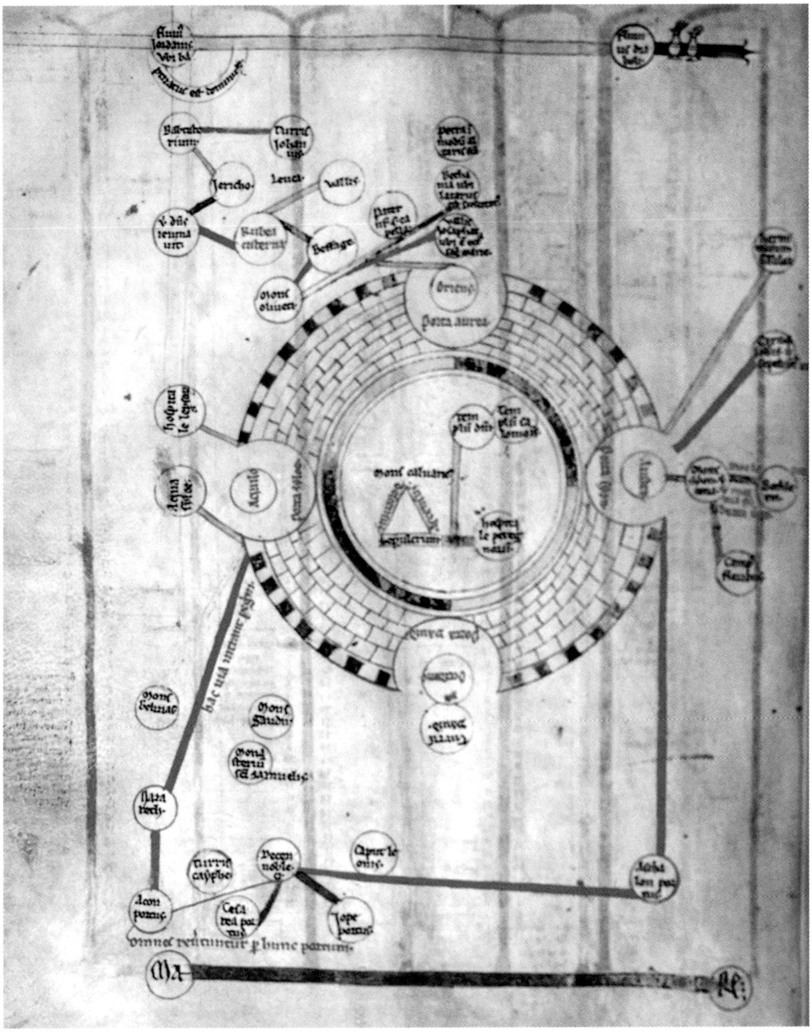

2 Map of Jerusalem, twelfth century.

circle, as with the earlier, ninth-century Carolingian examples such as the Valenciennes heavenly Jerusalem (illus. 38). This particular image shows the city's twelve gates, three situated on each opposing side of an encompassing walled circuit as scripture demanded, while the walled circuit itself is circular in form, depicted as twelve concentric rings, just like the order of the wider universe.[14] The symbolic significance of this imagined form is discussed by Frugoni, who follows Gousset in pointing to 'the choice of the circle as a symbol of perfection', and the idea that 'the circle is an image of the cosmos', with the city's gates positioned to form a cross, arranged to face the cardinal directions and point to the four corners of the world (a view first put forward some 50 years ago by Laveden).[15] Therefore the *imagined* circular form of the heavenly Jerusalem mirrored the imagined geometrical shape of the wider world, their shared geometries connecting one and other. The circle symbolized Jerusalem's cosmological importance.

Reconciling the heavenly city's two geometrical forms – the imagined circle and the scriptural square – also had a cosmological basis. Like the heavenly Jerusalem, the cosmos was also thought to be both circular (according to Neoplatonic cosmological sources) and square-shaped (according to Holy Scripture) (see below). To square the circle, illustrations of the city combined the geometries of the circle and square, a technique used in contemporary depictions of the medieval world as well as images of the holy city. For example, depictions of the circular Jerusalem, such as the Valenciennes manuscript, show opposing gates that form a quartered city, its cross-shape anchored to the cardinal points and the four quarters of a quadrate world. Another way that the imagined form of the heavenly city was geometrically connected to the wider Christian world is evident in the Triers Apocalypse manuscript where, as Frugoni has pointed out, the illustration shows a circle of walls but this time with emphasis on four of the wall's towers, which are coloured differently from the rest so as to form 'an ideal square', as scripture required (illus. 37).[16] Combining circle and square in this way not only reconciled the two forms of the heavenly city, it also connected city with cosmos. By having shared forms – the circle and the square – the heavenly city and the wider cosmos were mirroring each other; their imagery squaring the imagined circular cosmological form of the celestial city with how it and the world at large are both described in scripture, the word of God. These images of a geometrical heavenly Jerusalem, and the forms given to the celestial city, were also superimposed onto *earthly* cities, including Jerusalem itself, so connecting the metaphorical, imagined 'city' of God with the earthly 'cities of man'.

Depicting the holy city

Like its heavenly archetype, the earthly Jerusalem was also often depicted as circular in form. In Crusader descriptions of the city of the twelfth century it has a circle of walls pierced by gates positioned at three of the four cardinal points, with streets arranged to form a

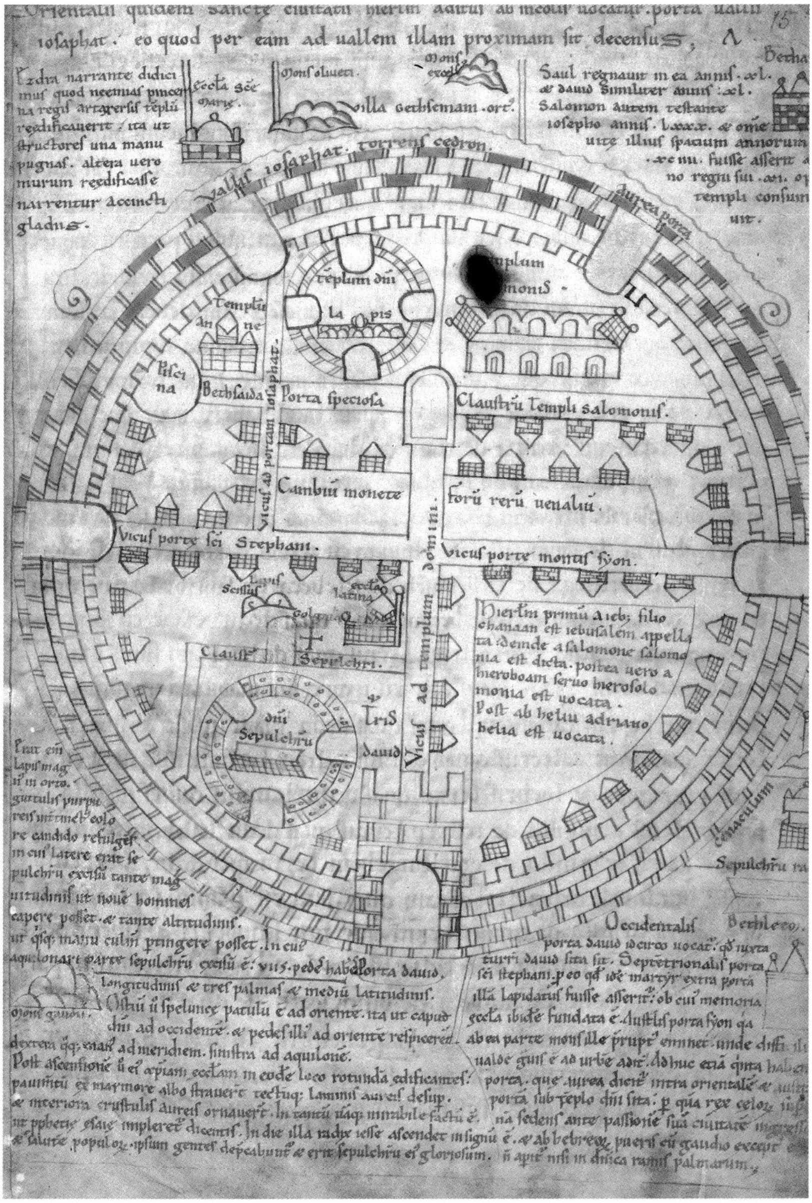

3 Map of Jerusalem, thirteenth century.

cross-shape and various landmark buildings indicated, such as the *Templum Domini* (illus. 2, 3 and 41).[17] This idea of the earthly Jerusalem as itself a reflection of the heavenly city gained currency in the context of the Crusader withdrawal from the city. Its 'loss' in the late twelfth century from Latin Christendom, and the desire to fight for and defend the city 'against the Infidel', is clear in a twelfth-century Crusader map of Jerusalem complete with fighting knights outside ('beneath') the circular-walled city with its cross of streets (illus. 41).[18]

While not all medieval images of the earthly Jerusalem used the same circular form as the celestial city – the thirteenth-century map of Jerusalem drawn by Matthew Paris for his *Chronica Majora* shows the city as square-shaped, for example – its circular form generally did predominate, as it did in images of the heavenly Jerusalem.[19] That the two Jerusalems shared a common form and layout is therefore clear. Their characteristics point to Jerusalem as an archetype, an idealized urban model visualized in abstract geometrical form.[20] As Alexander notes, 'in many representations of Jerusalem all that is required are walls', depicted as a circle- or square-walled circuit, 'surrounding buildings to signify [the] "city".'[21] These shared forms of the two Jerusalems circulated around the medieval Christian world, ending up far away from Jerusalem itself in cities such as London, Paris, Munich, Stuttgart and Florence.[22] Their impact on the minds of those who saw them can now only be estimated, yet something of their influence can be gauged by how this imagery of a circular Jerusalem took form in representations of other, European, towns and cities in the Latin West.

Medieval maps of Europe's towns and cities have yet to be brought together in a critical study.[23] Stylized urban representations of all sorts – not just 'maps' but other images, too – tend towards showing the city as having a circular or square form, seemingly deliberately and self-consciously imitating the forms of Jerusalem with its cosmological symbolism. Some medieval towns and cities really were geometrical in shape.[24] In those comparatively rare cases where urban forms are described or depicted by contemporaries, they were represented in such a way that over-emphasized beyond reality either the roundness or rectilinearity of their shape. One such example is a late-medieval representation of the city of Bristol (illus. 42). This appeared in a mayoral register begun by Robert Ricart, the town clerk, in the late fifteenth century, and with its circle of walls and four streets arranged in a cross-shape terminating at four gates, the image of Bristol has more than a superficial resemblance to those of Jerusalem.[25] However, it appears long after Crusader images of Jerusalem were being composed and one explanation of this is that 'Ricart was trying to depict what he thought Bristol looked like when it was founded', some five or six hundred years before.[26] This may well be so, for the earliest part of the city dates to the ninth century at least and was contained by defences that were roughly circular in shape with streets meeting to form a cross.[27]

Crusader images of Jerusalem drawn back in the twelfth century were also evidently circulating much later, as is evidenced by a fourteenth-century copy of an original of *c.* 1180 held by the Arnamagnæan Commission in Copenhagen,[28] and so equally the Bristol image may symbolize something more, echoing in its form the heavenly and earthly Jerusalems. The circular form of Bristol is also apparent in medieval 'maps' of other cities of about the same date, such as those of Vienna and Bratislava, with their chains of circular defences and streetless interiors which show some similarity with the toothed-circular defences of Jerusalem depicted in the Copenhagen and Florence maps, for example.[29] Then there is the rare example of a medieval plan of a town which apparently shows it in the course of being built: the plan of Talamone harbour in Italy of 1306, which has the town encompassed by a circle of walls but an orthogonal layout of streets inside, forming a cross-plan, so pointing once more to the circle and square being brought together in heavenly imitation.[30]

The forms of towns and cities on medieval maps thus reveal clear similarities with the Jerusalem archetype. As well as these cartographic representations of towns and cities, similarities can be found in other urban imagery. For example, images of cities are frequently shown on the wax seals that municipalities appended to their civic and corporate charters that endorsed their legal and constitutional urban status.[31] These seals were thus icons of civic status and authority, a self-image of the town or city as it was conceived by the corporate institution which ruled over the activities of those within its jurisdiction. Little attention has been given to the symbolism of these urban images, and yet they say so much about how the city was understood – how it was imagined – by those who lived in it.[32]

The earliest surviving English municipal seal belongs to a charter of 1191 for Oxford, a place with urban origins stretching back to the time of Alfred the Great.[33] The charter was 'issued by the citizens as a corporate body', and its seal, being 'the common seal of the citizens of the city of Oxford', 'depicts a complete walled city' within whose walls 'are three cylindrical towers'.[34] Some might argue the circle of walls on the seal is simply there because the seal itself is circular, but in Christian thinking 'the circular format is itself mystical since it reproduces the form of the world *in parvo*.'[35] Indeed,

4 Borough seals of Shrewsbury (*top*, fifteenth century) and Worcester (*below*, thirteenth century).

most English medieval municipal seals are circular in shape and thus analogous to ancient and medieval coinage; or else occasionally they are mandorla-shaped.[36] Both shapes were also used to depict the medieval world as well as cities (illus. 4 and 36). The circle was widely understood as an image of both city and cosmos, as we have seen, and the mandorla shape appears in Hildegard of Bingen's image of the universe, for example, as well as in an oval-shaped map of Rome from the early fourteenth century.[37]

This dualism between the shape of the world and the form of municipal seals is surely significant. Invoking this cosmological image – a circular or mandorla form like a map of the world – spoke 'of a pretended spiritual and temporal authority', reifying and legitimizing the authority and jurisdiction of the civic body (see chapter Five).[38] The circle of walls depicting the city on so many of these municipal seals thus echoed the image of Jerusalem at the centre of the world, imitating at the same time the heavenly city and the spiritual protection that it and its walls offered the inhabitants of the earthly city below.

Describing the city

How imagined forms of the heavenly city were being related to earthly cities emerges in written descriptions as well as visual urban imagery. Here again idealized images of Jerusalem, and the world it symbolized, were drawn upon by contemporaries who sought to describe their local and familiar urban landscapes. Their accounts reveal the imagined parallels between their own city and the city above: the heavenly Jerusalem; and the city at the centre: the earthly Jerusalem. By no means is there a wide abundance of such accounts,[39] but some have survived and make for interesting reading in the context of Christian cosmological symbolism. This is particularly evident in two descriptions of cities discussed here, Chester in north-west England, and Padua in northern Italy. In both cases their authors can be seen to have interpreted the forms of their localities in ways that connected them, and their city, with the holy city and the wider world that it represented.

De Laude Cestrie is a description in praise of Chester.[40] It was written in *c.* 1195 by an incumbent of St Werburgh's abbey named Lucian, who had been educated locally at the Collegiate Church of St John the Baptist in the city.[41] A particular flowering of written 'realistic observation' of the natural and human world was taking place in England at this time and Lucian's laudatory description of his home city forms a part of that tradition, sharing common influences of Classical works – in his case the writing of Virgil and Ovid, among others – as well as a 'spiritual love', as evidenced by his use of scripture and, of particular note here, by the allegories that he reads in and from the city's form.[42] The description appears to have 'originated as a series of sermons intended to please and edify the citizens of Chester',[43] a point worth bearing in mind for it would suggest that Lucian's reading of Chester's urban landscape reached an audience well beyond the local confines of cloister and court. This is what Lucian says about his city and its shape:

> [Chester] having four gates to the four winds, looks on the east to India, on the west to Ireland, on the north to greater Normandy [Norway] and on the south to Wales . . . There are two excellent straight streets in the form of the Blessed Cross, which through their meeting and crossing themselves, then make four out of two, their heads ending in four gates . . . [and] in the middle of the city, in a position equal for all, [God] willed there to be a market for the sale of goods . . . Now if anyone standing in the middle of the market turns his face to the east, according to the positions of the churches, he finds John the forerunner of the Lord to the east, Peter the Apostle to the west, Werburgh the Virgin to the north, and Archangel Michael to the south. Nothing is more true than that Scripture, 'I have set watchmen upon thy walls, O Jerusalem' [Isaiah 57:6] . . . So behold our city, as it was predicted, entrusted to the holy guardians as it were in fourfold manner. From the east the mercy of the forerunner of the Lord supports it, from the west the power of the doorkeeper of Heaven, to the north the watchful beauty of the virgin, and to the south the wonderful splendour of the angel.[44]

The city's four main streets that Lucian describes were already centuries old by 1195, possibly inheriting their alignments from Chester's Roman layout, and the streets were not in reality perfectly straight.[45] Lucian was therefore imagining the idealized form of the city's streets, and what was important to him was the symbolism that this cross-plan form conveyed. Through praising the urban landscape of Chester, Lucian is able to weave the city's form into a broader Christian allegory, reading from the city the symbolic significance of its landscape and at the same time imbuing it with symbolism from scripture and Classical works for his audience.

Not only is Lucian's Chester cross-shaped but its four streets are orientated to the four corners of the world, each with its own particular association and each pointing to spiritual anchors of the Christian world – the apostles and their saints. And the streets are not simply cross-shaped, for 'through their meeting and crossing themselves' the streets also imitate the devotional act of marking one's body with the sign of the cross. The cross-plan of the four streets and the city's gates orientated to the four cardinal points is an arrangement that has striking similarities to contemporary depictions of the earthly and heavenly Jerusalems. Indeed this (imagined) orientation and morphology of Chester is explained by Lucian through reference to scripture; 'as it was predicted, entrusted to the holy guardians as it were in fourfold manner'. His wording here evokes the 'four-square' quadrilateral form of the heavenly Jerusalem described in Revelation (illus. 39).[46]

In drawing parallels between his own earthly city and the celestial city, the heavenly Jerusalem, Lucian uses their shared forms as a heuristic device. Through its urban layout, Lucian was connecting the city of Chester with Christ and the wider cosmos. He is able to do this by interpreting the city's form as an imitation of the heavenly city, which in turn stands not only as a symbol of the Christian world – orientated, quartered and anchored – but also its foundation, its creation and destiny as the work of God, 'as it was predicted' from Holy Scripture. Moreover Lucian acknowledges God's role in making Chester itself, in placing the market at its centre for example, 'in a position equal for all', and also by providing other praiseworthy features, such as 'the rich and beautiful river beneath the city walls', placed there 'by the favour of God'.[47] In the form of Chester, then, Lucian's symbolic reading reflects his Christian understanding of the wider cosmos. One can only guess what kinds of texts he would have encountered as a student of the local collegiate church, but his learning of the *quadrivium* may well have included Neoplatonic works on cosmography and natural philosophy – perhaps extracts of Boethius or Calcidius – that likewise drew parallels that connected city and cosmos.[48]

The second of the two descriptions of cities that provides us with a glimpse of how urban forms were understood in the Middle Ages is later in date than Lucian's of Chester. It is likewise a laudatory piece, and concerns the city of Padua. Written in *c.* 1318 by a local judge named Giovanni da Nono, the description 'comes in the form of a vision given by an angel to comfort Egidius, the defeated legendary king of ancient Padua.'[49] Hyde noted that 'Da Nono puts his *descriptio* into the mouth of an angel who prophesies the future appearance of the city to console King Egidius of Padua after the destruction of the city by Attila the Hun.'[50] The description thus has cosmogenic resonances. These derive from scripture concerning fall and resurrection, prophecy and salvation. There is also perhaps an Augustinian influence, in that the struggle over Padua has echoes of that existing between the 'city of God' and the 'earthly city' – the Christian and pagan 'cities'.[51]

The vision with which the angel comforts Egidius makes reference to the city's form, and as with Lucian's account of Chester there appears to be a deliberate attempt by da Nono to use urban form in order to connect the city with a broader symbolic cosmology and cosmogony. Da Nono writes:

> I asked the angel in what form the city of Padua would be built. He told me: 'The fine wall of the city built by your Paduans will . . . curve round for a mile like a horse-shoe . . . and the water of the Bacchliglione and 'Tusena' rivers will flow round it . . . The Paduans will place four royal gates in the wall. The first will be called the gate of the mill bridge (Pontemolino) . . . The bridge at this gate will exceed the other bridges of the city in beauty . . . The second gate, towards the west, will be called the gate of San Giovanni delle navi . . . The third gate, to the south, will be called the Gate of the Toricelli, for in this part of the city will be built many more towers than in any other part . . . The fourth gate will be called the Gate of the Altinate bridge . . . From here the road will go to the Porto Ognissanti, from where boats will go to Venice . . . There will also be another fifteen minor gates . . .' [these gates are then each described in turn].[52]

Da Nono's description clearly places special emphasis on the four royal gates of the city, each placed at one of the four cardinal points.[53] Once again this appears to be an indication that he is drawing parallels between his city and the heavenly Jerusalem, with its gates on all four sides. The outline of Padua is also marked by a curved 'fine wall'. Perhaps most significant of all is his statement that the outline of the city was ordained from on high from an angel, one of God's own messengers in the celestial hierarchy, and that Egidius' image of Padua came to him in the form of a vision, just as in Revelation the heavenly Jerusalem is revealed to John as a mystical vision, a city of salvation descending from God (illus. 37). With da Nono's description of Padua's imagined mythical origins, such parallels with the heavenly city and all that it stood for, cosmologically and also cosmogenically, were surely being deliberately played up, for as Hyde has pointed out, 'while most description writers intended to praise and boast, da Nono's purpose was to warn,'[54] and in this case, as with the warnings in Revelation, and as developed in Augustine's *City of God*, the stakes were high for those who chose to ignore these prophecies.

With both Lucian's and da Nono's descriptions, then, there is evidence that urban forms – the imagined physical shape of a city – had meaning. In both cases features in the urban landscape were understood symbolically through the connotations they had, particularly their walls and gates. It is as if the authors were describing their respective cities to draw out parallels between them and the cities they knew from Scripture, cities that were to them redolent with significances to do with the history of the world – its cosmogony – their archetype being the earthly and heavenly Jerusalems.

As well as this, the cities that authors were imagining in their descriptions have an order and structure which replicated the order and structure of Jerusalem, as it was described in Scripture and as it was so often depicted in contemporary images of the period, such as those of the holy city with its circle of walls and opposing gates (illus. 38).[55] This was a city that not only had *cosmogenic* symbolism, as foretold by Scripture, but *cosmological* meaning too, as a microcosm of the wider world, evidenced in its form and structure, a scaled-down imitation of the form and structure of the cosmos. Through these two descriptions, therefore, we are able to see some indication that there were imagined cosmological and cosmogenic meanings associated with urban forms, at least in the later Middle Ages.

For the early Middle Ages there are fewer written descriptions of cities existing, and so the development of such symbolism is difficult to trace.[56] Nevertheless, one early example is a description of Dijon made by Gregory of Tours in his seventh-century *Historiarum*,[57] where once again we find a focus on city walls and cardinally positioned gates:

> It is a fortified place with very strong walls, built in the middle of a plain . . . Four gates face the four corners of the earth [*quattuor plagis mundi sunt positae*] and thirty three towers guard the walls. These towers are built of squared stones to a height of twenty feet and above these are courses of small stones. The

> total height of the walls comes to thirty feet and they have a thickness of fifteen feet. Why this place is not called a *civitas* I do not know.[58]

Here Gregory provides us with a particularly explicit statement acknowledging the cosmological orientation of the arrangement of Dijon's gates and walls, perhaps made all the more significant if we take into account Gregory's known connections with Cassiodorus, whose writings on cosmography at this time 'had become representative of the opinions of the more enlightened clergy' and 'enjoyed the respect and esteem of educators in the church schools'.[59]

Again, as with Lucian, it would seem highly likely that the authors of these descriptions were drawing not only upon scriptural works to inform their symbolic interpretations of urban forms but also other sources, including materials that pointed to a physical likeness between the city and cosmos. To understand this connection further requires us first to look more closely at medieval images of the world, at maps of the Christian cosmos, and examine the geometrical forms that it took, and second, to look at how geometry also provided a symbolic link between the city and world history. This will help to show how in their imaginings of both cosmology and cosmogony Christians of the Latin West were drawing close parallels between the city and their cosmos.

CHRISTIAN IMAGININGS OF THE MEDIEVAL WORLD

The circles and squares used to depict the heavenly and earthly city were present too in Christian imaginings of the medieval world. As conceptions of the cosmos, both circle and square were being used long before Carolingian cosmographers were drawing them in the ninth century.[60] In the tradition of Classical Greek cosmology the circle represented a spherical universe centred on a spherical stationary earth (illus. 43). Plato explains in the *Timaeus* how the universe came to be such a 'rounded spherical shape . . . a figure that has the greatest degree of completeness and uniformity', crafted by God.[61]

This Platonic view of the world held sway in the medieval Christian mind, passed along in a current of Neoplatonic thought through the writing of Macrobius, Calcidius, Boethius, Isidore of Seville, Eriugena and Gerbert, to achieve eventual intellectual dominance

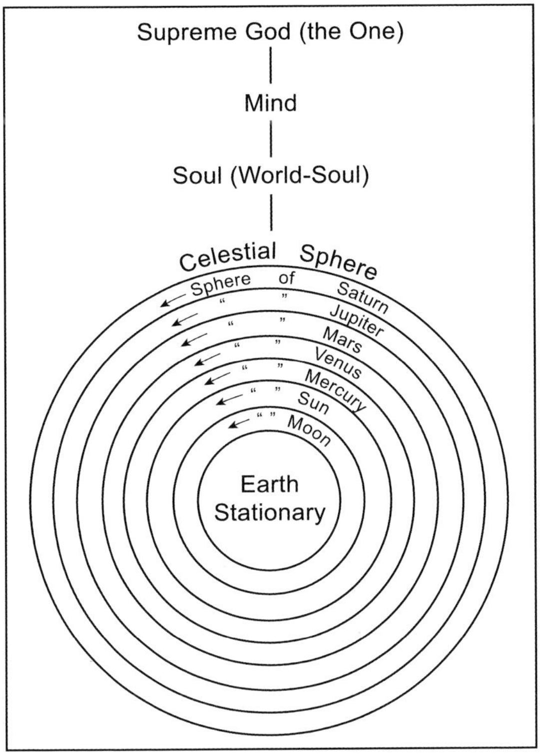

5 The Neoplatonic universe, diagram from Macrobius' *Commentary on the Dream of Scipio*.

in the eleventh and, more particularly, the twelfth century in the scholarship of William of Conches, Bernard of Chartres and other (Neo-)platonic Christian thinkers of this time, circulating through the many glosses on the *Timaeus* and copies of Calcidius' *Commentary*.[62] Even with the import of Aristotelian cosmology in the thirteenth century, the circular cosmos, 'with its succession of tightly nested concentric spheres', looked to all intents and purposes the same as it had in earlier Neoplatonic images of the universe (though of course they differed in their metaphysics).[63] The circle thus symbolized the geometrical form of the ordered cosmos, encompassing the celestial sphere, the planetary orbs and at the centre the earth itself.[64]

Platonic and Aristotelian images of the cosmos were set out time and again in the various drawings and diagrams accompanying cosmographies of the Middle Ages. In Macrobius' *Commentary on the Dream of Scipio*, for example, he used concentric circles to illustrate the spheres of the universe and motions of the planets, explaining how, 'since our eyes often open the way to the understanding of a

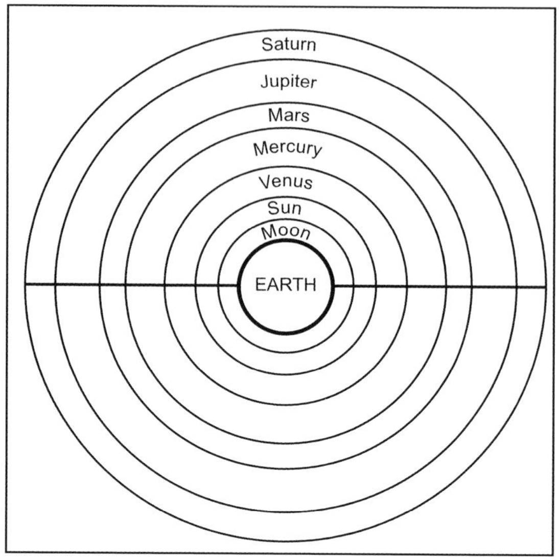

6 The celestial sphere: diagram from Calcidius' *Commentary on Plato's Timaeus*.

problem, it would be well to draw a diagram' (illus. 5).[65] Similarly, concentric circles depicted the cosmos in Calcidius' *Commentary on Plato's Timaeus*, with its circular-shaped earth and outlying planetary orbits (illus. 6).[66] Together these two influential early fifth-century texts were 'the most important source of Platonism in the Latin West in the Middle Ages', offering later medieval cosmographers a largely unified and unifying geometrical view of the terrestrial and heavenly worlds around them.[67]

The circular universe described in these Neoplatonic texts, and depicted in manuscripts as abstract diagrammatic images in their margins, was perpetuated through the *mappaemundi* that followed in later centuries; those medieval, and fundamentally Christian world-views that even today are still taken in vain by many modern commentators to be medieval 'maps'. But as Woodward and others have repeatedly made clear, these *mappaemundi* were symbolic representations, let us say imaginings, of the Christian world, and not charts by which to navigate the earth or maps used in the modern sense to mark out and define territories.[68] As symbolic imaginings of the Christian world, *mappaemundi* are richly iconographic and allegoric, and while there are differences in how features of the earth's surface are treated between *mappaemundi* 'types', the shapes used to encompass the globe are predominantly circular in form.[69]

Medieval conceptions of the world as revealed by the *mappaemundi* fused together Classical cosmology and Christian doctrine.

This fusion took place in the centuries prior to 1000 through the work of encyclopaedists, such as Isidore of Seville, the Venerable Bede and, later, Hildegard of Bingen, who sought to describe the visual appearance and form of the world.[70] A *mappamundi* accompanying the early seventh-century *Etymologies* of Isidore of Seville provided a model image for later *mappae*, featuring a circular-shaped earth with a tripartite division representing the three continents of Asia, Africa and Europe, the whole sitting at the centre of the circular Platonic universe.[71]

Mappaemundi such as the well-known 'Psalter' and 'Ebstorf' 'maps', both from the thirteenth century, and many other less well-known examples, perpetuated this particular 'geometric scheme' (illus. 36 and 44).[72] At the same time, images such as these also presented a view of the earth as a 'map' of Christian belief. The 'T' formed by the tripartite arrangement of the three continents symbolized the *tau* cross-shape;[73] while, as Kühnel observes, the earth depicted as 'a concentric entity' represents 'in one image the whole history of salvation, backed up by Christ himself embracing the disc', as in the 'Psalter' and 'Ebstorf' maps where 'Christ is at one and the same time the Crucified (suggested by the position of his head, hands and feet at the four cardinal points of the 'Ebstorf' map), the Resurrected (suggested by the censers held by the angels in the 'Psalter' map), the Creator (suggested by the way he holds the disc of the world . . .), and the Saviour of the end of days.'[74] By the twelfth and thirteenth centuries, then, 'a *mappamundi* could thus represent simultaneously the complete history of the Christian world: its creation, salvation, and final judgement',[75] and as such the circular form of the world came to symbolize the Christian universe and its past, present and future.

While the circle certainly dominated in medieval conceptions of the cosmos, it was by no means unique. The square also featured as a geometric form in Christian imaginings of the world, sometimes in combination with circles to represent the motion of the sun around the earth. A four-cornered, rectangular-shaped earth is to be found, for example, in computistic compendia dating from the ninth to the eleventh centuries.[76] In these the *mundus* is four-cornered, as it is described in the Bible and as it was seen by Cosmas Indicopleustes in his *Christian Topography* of the sixth century.[77] In one of these diagrams, an eleventh-century manuscript, a rectangular-shaped *mundus* encloses the three continents of Asia, Europe and Africa, while in another contemporary example the *mundus* is more square-shaped,

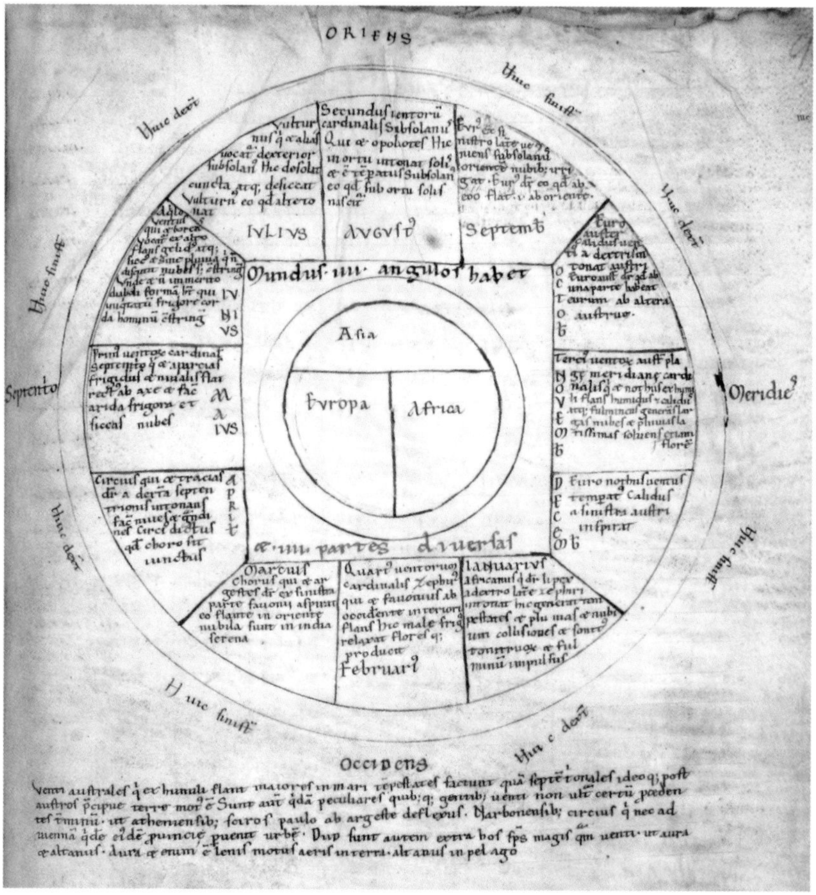

7 Eleventh-century image of the world.

and within it is shown a circular form of the earth, again with the tripartite division of the three continents, with 'a caption placed along the inner upper and lower sides of the square specif[ying] that the world has four angles and four different parts.'[78] (illus. 7) Here, then, the circle and square are combined, and brought together to simultaneously represent the two shapes of the world as they were in depicting the heavenly city. In Cologne Cathedral library a further example of *c.* 800, identified by Obrist, has a perfectly square-shaped earth, placed at the centre of concentric circles, with the whole scheme contained within an outer square (illus. 8).[79] In this image the two, apparently competing forms of the world – the world as rounded and as four-cornered – are reconciled. This was at the time that Rabanus Maurus, the German encyclopaedist, was writing about 'how circular and quadrate shapes could agree', using a Euclidean

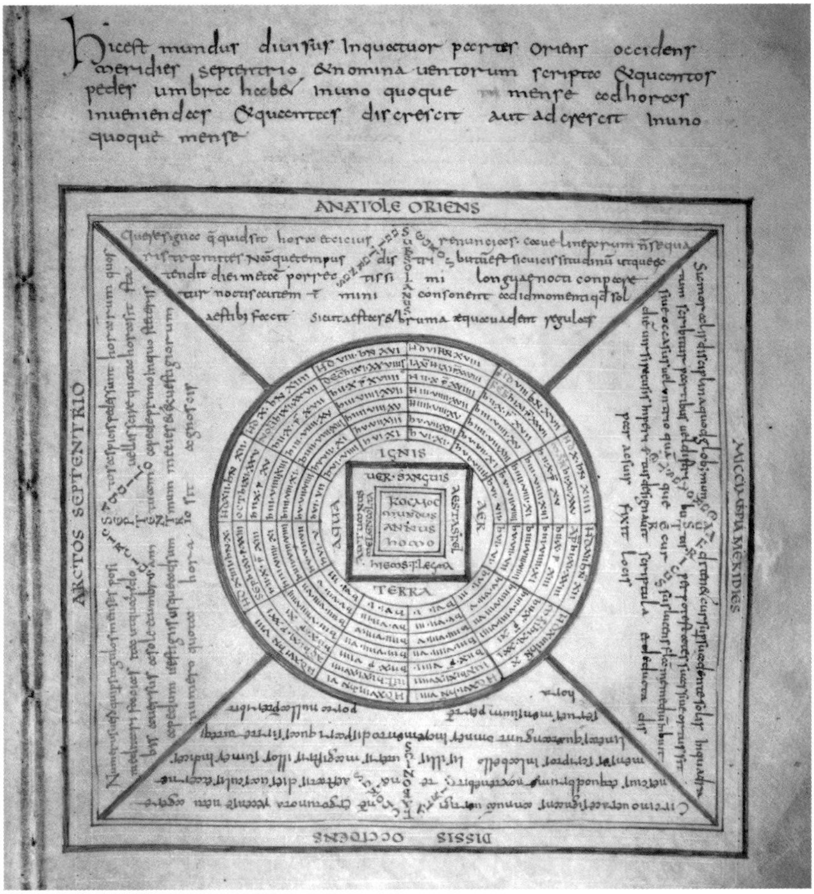

8 Image of the world, *c.* 800.

solution to 'squaring the circle', and so reconciling geometrically the *orbis quadratus*.[80]

A square-formed world not only had scriptural backing, but also a broader mysticism associated with the number four which reached back to Classical times and a Timean cosmology. The *Timaeus* relates how the soul of the world was formed by God from two strips 'placed crosswise at their middle points to form a shape like the letter X', which were then bent round to create 'two circles, one inner and one outer'.[81] Plato's description resonated through the Middle Ages, in part through Calcidius' commentary, but more particularly in the twelfth century in glosses and commentaries on the *Timaeus*, at a time when the world-soul was being discussed by the likes of Alan of Lille and Bernard of Silvester in their literary cosmographies.[82]

The eschatological significance of forming the world-soul from a cross-shape would surely not have been lost to those familiar with the Timean account.[83] Certainly, the number symbolism of the *Timaeus*, and the significance of the number four particularly, was taken up by Christian thinkers in their conceptualization of the structure of the world and its components.[84] The *Timaeus* describes the 'four constituents' of 'the body of the world' – air, fire, water and earth – and their proportional and geometrical relationships to one another, that when 'put all together were supposed to form a cube', and which, in subsequent Neoplatonic writings, such as Isidore of Seville's *De Rerum Natura*, came to be presented as 'a single-plan figure', a square.[85] One example of this, where the four elements are combined with a representation of the earth, is a diagram in a ninth-century manuscript of Bede's *De Natura Rerum* showing the relationship between 'the four cardinal directions (and the three continents), the four seasons, the four elements, and four material properties (hot, cold, wet, dry)' (illus. 45).[86] The four corners of the earth itself marked out the shape of the true cross, as one seventh-century Irish commentary on the Gospel of St Mark put it: 'what is the very appearance of the cross if not that of the quadrate form of the world?' (*ipsa species crucis quid est nisi forma quadrata mundi*).[87] The symbolic significance of the number four, a 'symbol of moral perfection', and its association with a square-shaped world, thus converged in the Christian mind through Neoplatonic medieval cosmographies with their images of a spherical universe.[88]

Through looking at medieval images of the cosmos, its geometrical structure, the circle and the square, something of their Christian meanings emerge. The medieval universe was fundamentally ordered, part of a 'celestial hierarchy',[89] as those such as Hugh of St Victor saw it:

> Their universe was a development, by way of Boethius' writings on arithmetic, of the Augustinian principle that God disposes all things in accordance with order and measure – a principle which combined the classical concept of the cosmos as *consentiens conspirans continuata cognatio* [harmonious and continuous relations] with the principle of a Divinity who is life, providence, and destiny.[90]

There was, then, an aesthetic to the 'geometric schema' of the cosmos that not only reflected God's beauty but existed in everything, everywhere, in all that he had created. Again, this is a Neoplatonic and Timean thread woven into and through the fabric of medieval Christian cosmology, especially that of the eleventh and twelfth centuries, when the influence of Platonism was at its peak in circles of medieval theology and philosophy, in the cathedral schools of Chartres and Laon, for example, and within a broader network of intellectual ties that bound together Christian thinkers in the Latin West.[91]

Recently, Kühnel has pointed out how 'Carolingian preoccupations with the shape of the Christian universe influenced not only the shape of "maps of the world", but also the representations of the Heavenly Jerusalem of Revelation, equally forcing them into strict geometrical schemes.'[92] This is seen by placing images of the cosmos, the earthly and heavenly Jerusalems and the world from the ninth to the thirteenth century side by side (illus. 38, 41, 43, 44). In medieval maps of the world, as well as in images of the city, the same geometrical schema are used, the circle and the square. The reason for this, Kühnel suggests, is that the 'city of Jerusalem, and especially its *loca sancta*, were absorbed into the geometrical perfection of the Christian universe', a conceptual dualing of both the heavenly and earthly worlds.[93]

What we can deduce from comparing images of the Christian world with contemporary depictions and descriptions of earthly and heavenly cities is that they share a sacred geometry. Particular shapes – circles and squares, the cross and cardinal axes – connected city and cosmos both imaginatively and materially. In images of city and cosmos, urban form and cosmological form were analogous, and their shared geometrical forms conveyed a common symbolic meaning. Both formed part of a hierarchy of concentrically ordered spaces from centre to edge, and both were marked with the sign of the cross, symbolizing Christ himself. These Christian urban and cosmological imaginings, with their common structuring 'geometric schema', were circulating at the same time as philosophers and theologians were drawing upon Neoplatonic and Aristotelian sources to elucidate and conceptualize the form and order of the cosmos (see Introduction). The same sacred geometries were also being used in contemporary accounts of the world's creation in recounting cosmogony, and again they forged a symbolic link between city and cosmos in the medieval imagination.

COSMOGONY AND THE CITY IMAGINED

In various iconic images, then, the medieval city was represented in ways that tied together city and cosmos. Their shared geometric forms in this imagery point to this conclusion. But while images may show the same forms, this alone might be doubted as evidence of a shared symbolism. The imagined city's shape reflected not only cosmological form but also cosmogony – of how the world came into being and how it will end – as told principally in Scripture by the books of Genesis and Revelation. Medieval sources describing the cosmogenic act also depict God as an architect forming the universe using geometers' tools, the compass and quadrant or set-square, which were also those that were used to form the city and give it its geometrical shape. The city thus represented a microcosm of the formation of the world as well as its form. Both derived from a Timean view of the world, and of creation, which contemporaries attempted to square-up with Scripture.

First, the circle of the *mappamundi* and the circular form of the city represented cosmological time as well as space. Thus, 'the spiritual history of the Christian world, from its Creation to the Last Judgement, with a sequence of divinely planned events in between, such as the Salvation by Jesus Christ, are all carefully portrayed – in more or less detail – on *mappaemundi*', most obviously as shown by the 'Ebstorf' map with the Body of Christ superimposed on the disc of the world, but also in the way that the author of the Hereford map saw 'his graphic work as a "history"' of the world.[94] So, too, did the circular form of these maps represent, allegorically, a wheel that was itself symbolic of the cyclical passage of time.[95] Isidore of Seville 'describes the earth as a "circle of lands" like a wheel' in his *Etymologies*, and saw the roundness of the world (*mundus*) also as a year (*annus*), as is evident in the *orbis quadratus* depictions of the universe of the eighth and ninth centuries (illus. 46).[96]

Later, in the twelfth and thirteenth centuries, world maps were also combined with an image of 'Fortuna's wheel', reinforcing the link between time and space, albeit through the activities of a pagan goddess, as Honorius of Autun asks rhetorically: 'what is this wheel? It is the glory of the world which is carried round in perpetual motion.'[97] A popular image in the medieval Latin West, Fortuna was mythologized in the Neoplatonic literary cosmographies of the twelfth century by Alan of Lille and others: 'her caprices always associated, at least implicitly, with the celestial forces of fate, the gods and the stars'.[98] The roundness of the medieval world, like Fortune's wheel, thus symbolized its movement as well as its eschatological history, as did the circular form of Jerusalem, 'the place of the end of days', the *axis mundi*, depicted, like the world around it, as a wheel – or, more pointedly, a toothed cog – located at the centre of a *machina mundi*.[99]

In Augustine's *City of God*, the city has a vital place in the Christian history of the world. The 'city' for him was at once both prophecy and salvation, as told by the Bible. The fall of Rome was thus forewarned in holy scripture, in Genesis, by the founding of Enoch, for both it and Rome had been built on fratricide; while the earthly city of Jerusalem foreshadows the coming of the heavenly Jerusalem, in the Final Judgement, as told by Revelation.[100]

For Augustine, then, the cities of Enoch, Rome and Jerusalem, as well as Babylon, were markers in the history of the world as revealed by scripture, while the 'City of God' itself – 'consisting of the good, angels as well as men' and contrasted by Augustine with 'the other of evil', the 'earthly city', the two of which 'are interwoven and intermixed in this era, and await separation at the last Judgement' – is a figurative 'city' of which 'we have longed to become citizens . . . with a love inspired by its founder'; a city, in short, of Christian salvation.[101] Such imagery took pictorial form in the later Middle Ages, when the city of God is 'seen in the shape of a parallelogram, sited on a mountain, and set in a circle' on which 'the figure of Christ is seated at the apex', the corner stone of this 'city', this world.[102] Augustine's 'city' was not only symbolic of the Christian history of the world, therefore, in the guise of the 'earthly city', but was also itself a metaphor for God's hand in its creation in his role as its 'founder', an idea deeply rooted in Timean cosmogony, where it was God 'who proceeded to fashion the whole corporeal world', as would an 'artist' or 'great artificer'.[103] Just as images of a circular city and cosmos symbolized a history of the Christian world, its cosmogony and its yearly cycle, so too was the cosmos imagined to be like the 'city' – whether the 'earthly city' or the 'city of God' – founded by its creator. Both of these images drew the city and cosmos closer together in the medieval Christian imagination.

Second, the cosmogenic symbolism of the city is evident in later medieval depictions of God as 'architect' of the universe, and here the circle and square imagery of both the city and cosmos are given additional meaning by God's use of an architect's instruments in the making of his universe, particularly the compass, or dividers, used to create a perfectly-drawn circle (illus. 47). In a fifteenth-century manuscript of Augustine's *City of God*, the creator is shown with compass in hand 'engaged in his handiwork', an image that earlier in the twelfth and thirteenth centuries was widely used in depicting the act of Creation in illustrations 'extant in psalters, *horae*, picture Bibles and other manuscript books'.[104]

In these 'Creation miniatures', as Friedman notes, 'the compass commonly appears in a symbolic or emblematic role', 'an emblem of the creation itself' based on 'a metaphoric conception of God as an architect or artisan forming the cosmos from unorganized matter'.[105] The Timean idea of God as the 'great artificer', reiterated in twelfth-century Neoplatonic cosmography by William of Conches, Thierry of Chartres and Clarembald of Arras, also appeared at around this time in Creation miniatures, with God depicted tracing out a circular outline of the world with his compass, as an architect would trace out a circle to make plans and designs for building a new cathedral or abbey, or city.[106] Furthermore, the compass is not just a tool for drawing circles but an instrument for measuring them as well. So the circular universe that God had created with his compass was what the geometers of the ancient and medieval world were measuring with theirs in their works on cosmography and cosmimetry.[107]

The compass provided a symbolic link between the city and the cosmos, for the circular world fashioned by God's compass was also the circle of the imagined city, each being formed in the same way. The earthly and heavenly city became, then, a microcosm – *microcosmus* – of the world at large, the two cities and the cosmos having been created by the work of an 'architect', as Philo of Alexandria made clear in his account of God's creation of the world:

> When a city is being founded to satisfy the soaring ambition of some king . . . there comes forward now and again some trained architect who . . . first sketches in his own mind well nigh all parts of the city that is to be wrought out . . . and like a good craftsman he begins to build the city of stones and timber, keeping his eye upon his pattern and making the visible and tangible objects correspond in each case to the incorporeal ideas.
>
> Just such must be our thoughts about God. We must suppose that, when He was minded to found the one great city, He conceived beforehand the model of its parts, and that

9 Life of St Offar, late fourteenth century.

> out of these . . . He brought to completion a world discernible only by the mind, and then, with that pattern, the world which our senses perceive.[108]

God the 'architect', who 'was minded to found the one great city' – the cosmos – is thus symbolically likened to the architect that 'builds the city of stones and timber', the earthly city. Both created something from nothing, as it were, as later sources show so clearly, using compass and square in their work (illus. 9, 47 and 48).[109]

Similar architectural metaphors were of course to be found in the Bible itself, as well as in Neoplatonic works, while later in the twelfth and thirteenth centuries, Hugh of St Victor, Richard of St Victor and Roger Bacon, and others of their kind, realized 'the importance of geometry as an aid to understanding scripture.'[110] What makes this interplay of 'city' and 'cosmos' especially significant is the way that it took shape throughout the Middle Ages in the imagery of *both* city and cosmos. A Christian, cosmological symbolism of circle and square was plain, therefore, to all who used these geometrical forms to picture the world's physical shape. The tools from which God had made it – the compass and square – were likewise those that were used to create the forms of both the heavenly and earthly city, themselves a symbol of the cosmos. Not only were the compass and the square imbricated in the construction of this sacred cosmogony, so too were other instruments, such as the circular-shaped astrolabe, used in cosmimetry as well as altimetry and planimetry, the surveyor's quadrant, which combined circle with square to measure angles as well as straight-line distances, and the 'geometric square', similar in function to the astrolabe – all were geometrical instruments used for creating and measuring the city and the cosmos in the Middle Ages.[111]

To sum up, conceptions of city and cosmos, and the imagined connections that existed between them, took geometrical form in descriptions and depictions of the heavenly and earthly city, and in cosmological imagery as well. These 'urban mappings' were informed in part by what medieval Christians knew from Scriptures about the world and the place of Jerusalem within it, and the divine order it represented, and in part by the ideas of their contemporaries, Classical commentators and philosophers who saw the city as both macro- and microcosm. It was in their shared sacred geometrical forms that city and cosmos were linked together in the Christian imagination, coupled with a belief in medieval minds that these symbolic geometries expressed the work of God, both as cosmocrator and creator. Through a sacred geometry of circles and squares the imagined earthly and heavenly 'city' became in the medieval mind an image of God's universe, both an image of its form (its cosmology) and its formation (its cosmogony).

TWO

Urban Forms

URBAN LANDSCAPES AND SYMBOLIC FORMS

Idealized urban images and descriptions were circulating at the time that new urban landscapes were being physically formed in the Middle Ages. In some cases these urban landscapes also took geometrical form, raising the question as to whether sacred geometries and cosmological symbolism were being intentionally written into them. While physical aspects of medieval towns and cities have in fact long been a subject for academic study, what has so far escaped attention is how these built forms of medieval urban landscapes were inscribed with symbolism through the particular shapes that were used to set them out on the ground.[1] To begin with here the focus is on the presence of rectilinear and curvilinear layouts in medieval urban landscapes, and their square and circular forms. These geometrical shapes, it is argued, were chosen deliberately by those creating new urban landscapes, but not simply for pragmatic or utilitarian reasons, rather to convey a symbolic form that was itself rooted in sacred geometries common to both city and cosmos.

Since virtually no visual representations exist to show what physical form medieval urban landscapes took at the time of their making, their original ground-plans need to be reconstructed combining analyses of surviving patterns of medieval streets and plots with available historical and archaeological material.[2] These built forms – the structuring framework of street and plot patterns that make up European town-plans – are a monument to medieval urban culture and civic industriousness. In general it is those smaller medieval towns whose plans have attracted most careful study, especially those having more regular and rectangular-shaped built forms, since these in particular are seen to be exemplary of 'medieval town planning'.[3]

Four-square geometries and their symbolism

Urban landscapes with distinctly rectilinear forms continue to receive more attention than others. They are almost invariably associated with urban foundations, 'new towns' that were being formed all across Europe during the Middle Ages, more especially in those regions undergoing settlement and colonization.[4] Although certain groups of these towns have had their plans particularly well studied, little consideration is given to what these rectilinear forms might have meant to those who created them at the time.[5] The *burhs* of Anglo-Saxon England, created during the ninth and tenth centuries, are a case in point, having had fairly thorough treatment of their urban topographies.

At the end of the ninth century England was divided politically between warring Danish and Saxon dynasties. *Burhs* – principally fortified places – provided defensive structures for both sides of the territorial divide, being established both in the eastern Danelaw area during the tenth century, and rather earlier in southern England, in Wessex, under Alfred the Great (illus. 10).[6] The *burhs* were in some cases new towns placed on greenfield sites, though some use was also made of a former Roman *civitas*, as with Winchester and Exeter.[7] With the new urban *burhs* of Alfred's reign, places such as

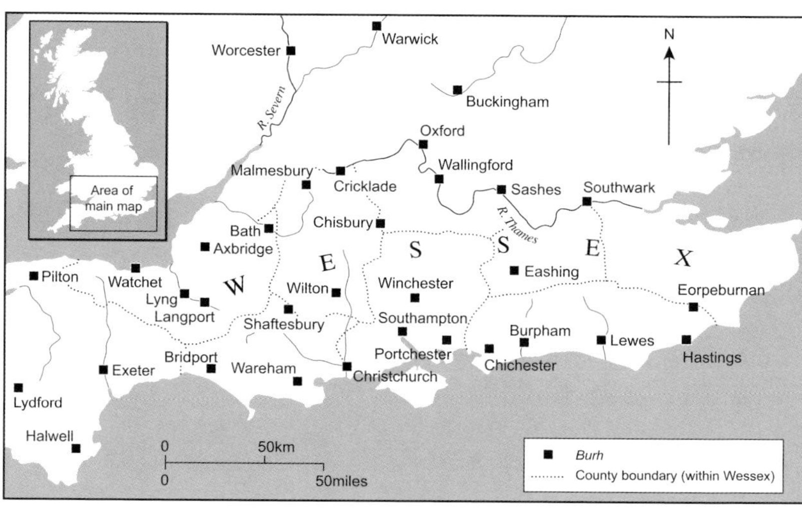

10 The *burhs* of Anglo-Saxon southern England.

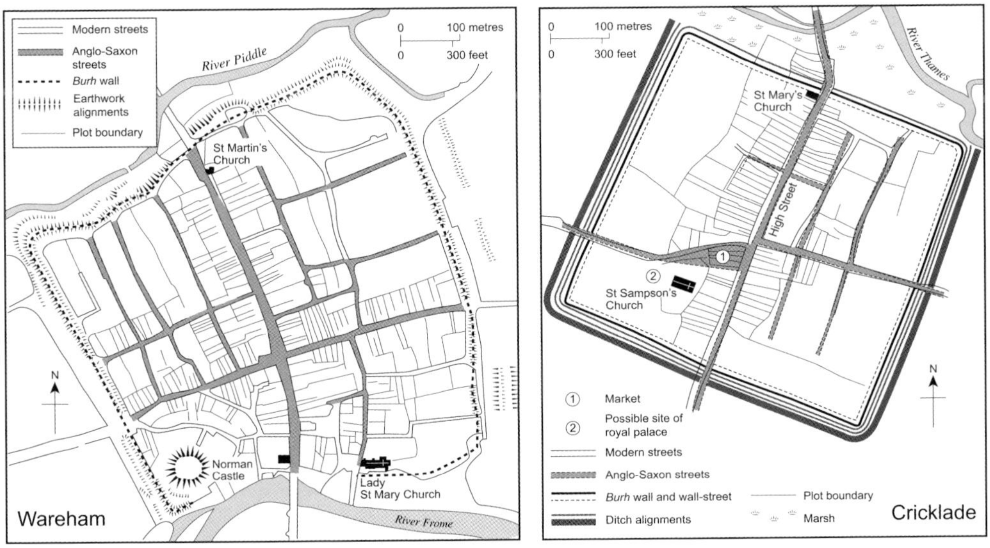

11 The layout of King Alfred's Wessex *burhs* compared: Wareham and Cricklade.

Cricklade, Wallingford, Oxford and Wareham, rectilinear forms are particularly clear in street and plot patterns contained within a squarish-shaped defensive enclosure, in most cases set alongside a watercourse (illus. 11).[8]

As 'fortifications commanded by the king', the square-shaped defences of Alfred's *burhs* were integral to, and defining of, the layout of the new towns, giving them a quadrate form that was replicated by the street and plot forms within the defences.[9] The streets and plots themselves are quadrilateral, with two streets meeting in a cross shape providing the principal axes through the towns, and elongated street-blocks making up each of the four quarters. The overall impression is one of spatial order, as is also the case with those *burhs* that were sited within reused Roman defences, as at Chichester.[10] Crummy and Haslam have attempted to quantify the regularity in the layouts of the Anglo-Saxon *burhs* in southern England by measuring the distances between streets and calculating the dimensions of street blocks.[11] Their work reveals a strong uniformity between *burh* layouts both in terms of the unit of measure used to create them – the sixteen-and-a-half foot pole – and in the positioning of their internal streets and street-blocks.[12] Indeed, while there are elements of rectilinearity in the plans of the five Danish 'boroughs' established in eastern parts of England, at Stamford and at Nottingham for example, they are noticeably less quadrangular in form compared to their Saxon counterparts in Wessex (illus. 11 and 12).[13] This suggests that the

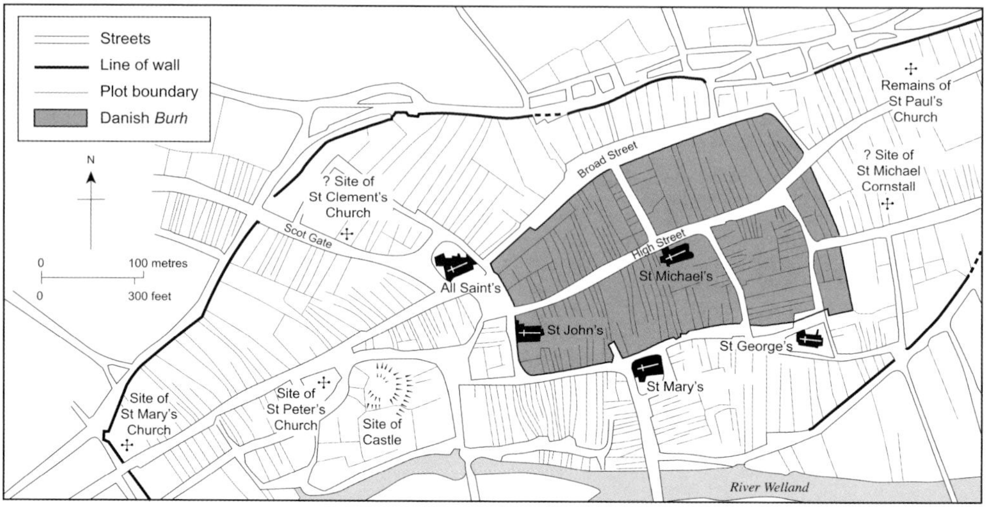

12 Danish *burh* of Stamford.

Alfredian *burhs* are a morphologically distinct group of towns, each having certain design characteristics in common with others of the same origins.

How far Alfred had himself been involved in choosing a quadrilateral form for the Wessex *burhs* is open to question, though judging from what Asser, the king's biographer, said about 'the cities and towns he restored, and the others, which he built where none had been before', it is possible that Alfred did have some direct influence.[14] The *burhs* were royal towns, after all. If a model for Alfred's *burhs* is to be sought it may have come from Classical precedents, either through surviving remains of Roman *castra* in England, or conceivably even the eastern Mediterranean, and in this regard it might be that Alfred was being influenced by Charlemagne.[15] Their quadrate layout might also have been inspired by the imagined quartered form of the heavenly Jerusalem (illus. 39).

Images of the celestial city were being produced in Carolingian centres of learning at this time, in places that Alfred had firsthand experience of.[16] One of these centres was Aachen, where Charlemagne the Great had his principal palace, and it seems noteworthy that there are morphological similarities to be seen when Carolingian Aachen is compared with the contemporary layout of London, a city which Alfred 'restored' following Danish occupation of the city in the 870s.[17] But even if Alfred's Carolingian connections are left aside, and likewise those contemporary images of the celestial Jerusalem circulating at this time, a squared-form model could be found in

Scripture alone. That Alfred was thinking about the heavenly city in a Neoplatonic cosmological sense is made clear through his translation of Boethius' *Consolations*, undertaken between 887 and 899, in which he transposes the 'city-state' of the Latin text for Jerusalem.[18] Not only were Neoplatonic ideas of the city and cosmos familiar to him then – through reading Boethius and also Augustine's *Soliloquies* – Alfred was using them to consider his own divine authority.[19] All in all the rectilinear forms of Alfred's *burhs* point to a significance hitherto unappreciated by modern scholars, despite the attention that this group of new towns has received: the quartered layout of his *burh* being perhaps modelled on the square form of the heavenly city itself. Alfred's towns thus mirrored on earth the imagined form of the city-cosmos which Jerusalem represented – so affirming Alfred's kingship as divinely ordained by God and his place in the cosmic order (see chapter Three).[20]

By no means are Alfred's four-squared *burh* layouts unique. There are similar examples all over medieval Europe. Two particular regions where new towns commonly had this form are east-central Europe, in parts of modern-day Poland and the Czech and Slovak republics, as well as southwestern France. The new towns of east-central

13 Vianne (southwest France), an English *bastide* town chartered in 1287. View through the Porte Notre-Dame, looking west.

Europe were being created by *locatores*, agents acting on the part of aristocratic lords to deliberately promote Germanic settlement and colonization, while in southwest France contested areas of Gascony were populated by new towns (*bastides*) created by both the French and the English Crown through local lords, seneschals and members of the royal household employed by the king (illus. 13).[21] In both regions, although the towns are later in date than Alfred's *burhs*, they share a layout that is rectilinear and quadrilateral in form. This recurring concern to make new towns from a square shape is surely not simply a coincidence? The square, as we have seen, was a geometrical form used to symbolize stability, a characteristic that was no doubt desired for towns being placed in what were, in the case of east-central Europe and southwest France, both contested frontier regions that were politically volatile because of dynastic rivalries and territorial claims.[22] The cosmological symbolism of the square may thus have been a reason for using it to shape these frontier towns, coupled with the image of the heavenly Jerusalem with its promise of salvation and a better world to come.[23] Such interpretations of the built forms of these new towns tend to be overlooked, however, almost without exception.

The *bastides* of southwestern France are a second group of well-studied medieval 'new towns', again noted especially for their regular layout and rectilinear grid-plan forms of streets and plots.[24] Certain of these, such as Aigues-Mortes and Monpazier, are frequently cited as *bastides par excellence* with their internal chequer-board plans, overall quadrilateral shapes and fortified outlines.[25] Typically, the characteristic geometrical *bastide* plan is interpreted from a 'functional' perspective by historians.[26] For example, the practical economic and administrative advantage afforded by having new towns of rectilinear form is emphasized in a discussion of the *bastides* by Randolph:

> The utilization of a standard urban model expedited the construction process, by giving the sovereign's representative in the field – usually the seneschal – a set of operating assumptions and therefore the liberty to work without recourse to the bureaucratic hierarchy; the grid plan [of the towns] would impose order, facilitating tax collection and administration – the inhabitants would be administratively visible ...; the block geometry facilitated growth, providing a system of logical urban expansion. [27]

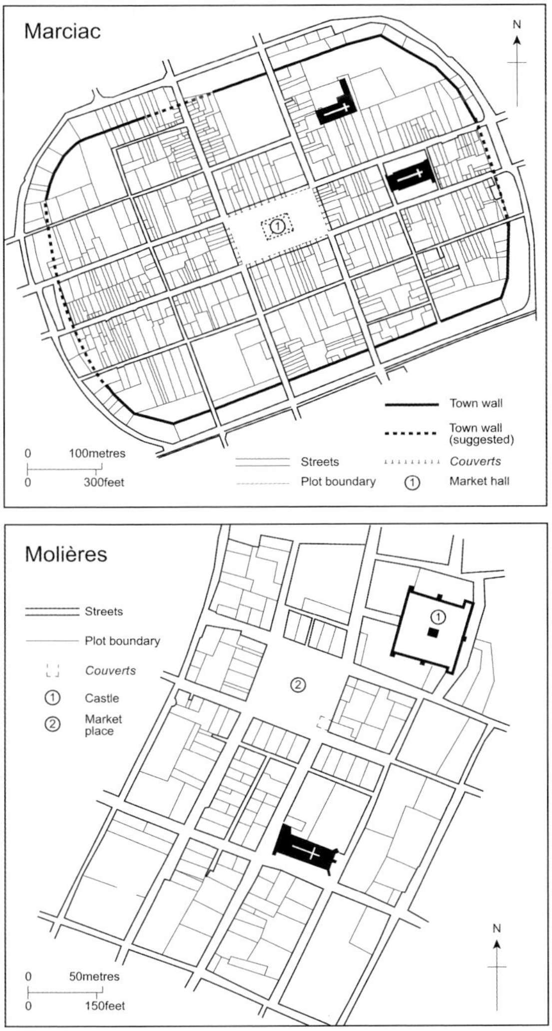

14 New towns in southwest France; the *bastides* of Marciac and Molières.

For Randolph, then, as is true for others who stress practical expediency as the basis for *bastides*' gridded forms and layouts,[28] there is apparently no space in their plans for anything but utility. It is as if the *bastides* were formed as part of some bureaucratic zeal to serve a modern nation state rather than created in a world where, as Mâle had put it, 'all forms are endowed with spirit'.[29] This was surely more how the *bastides*' built forms were seen at the time, in the thirteenth century?

Perhaps because some of the *bastides* look as if they are made up of rectilinear street-blocks in a grid-plan simply cast over the

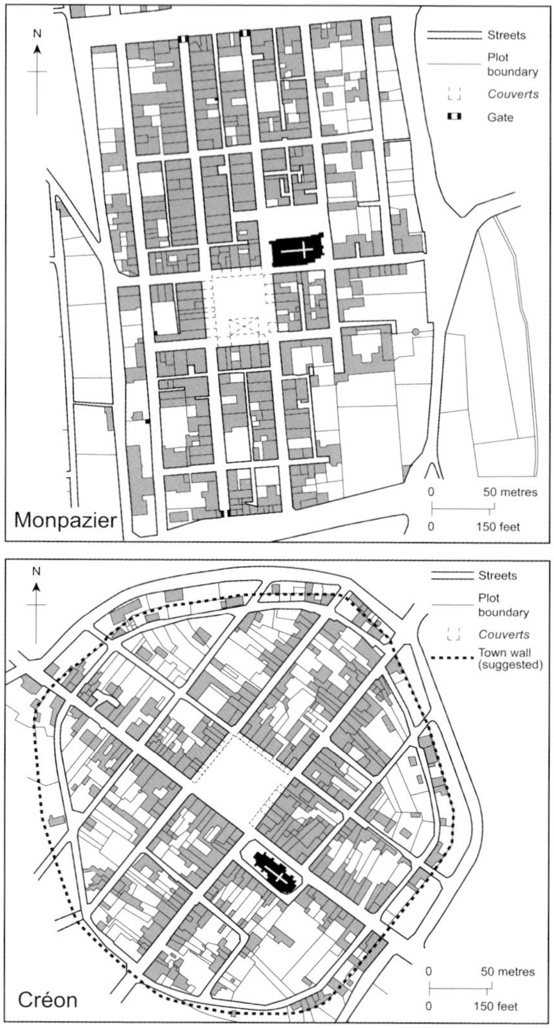

15 New towns in southwest France; the *bastides* of Monpazier and Créon.

landscape as a kind of loose net, there is a tendency to view them as a single type.[30] Rectilinearity is undoubtedly a common morphological trait of the *bastides* of southwest France, but recent detailed and comparative study of their forms has revealed subtle variations in their morphological make-up.[31] Variations exist in the linearity of street-blocks or the position of the church, for example.[32] Despite such differences, what characterizes most *bastide* plans is a large open and square-shaped *place*,[33] often located at the very centre with two axial streets entering at right angles at each of its four corners (illus. 14, 15, 16). In this sense, the *bastide* plans show similarities to

16 Monpazier, the *place* and its *couverts*.

the layouts of some east-central European new towns.[34] The *place* was of course the *bastide*'s commercial focus but it has also been suggested that it played a role in determining their overall rectilinear form. A case in point is Grenade-sur-Garonne, a *bastide* created by Eustache de Beaumarchais in 1291, where Bucher has shown that the diagonal distance taken across the square-shaped marketplace was used as a basis to generate the dimensions of the town's other street blocks (illus. 17).[35]

Not all the *bastides* provide such a neat example of this as Grenade, but Randolph nevertheless suggests that generally 'the

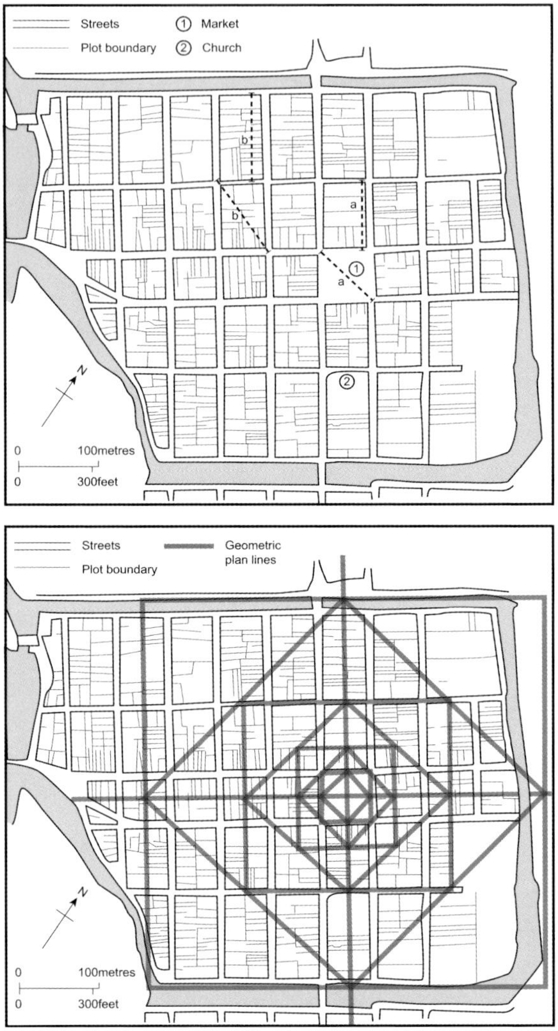

17 The 'design geometry' of Grenade-sur-Garonne (France): the diagonal distance 'a' (across the central square) equals the 'auron' distance 'a' (of the elongated street-block) (top); use of 'rotating squares' to establish the town's plan (below).

marketplace, as the basic generative unit in the block geometry of the *bastides*, played a fundamental role in their regular, yet flexible, expansion', each street-block in effect being part of a modular system that could be added to as and when desired.[36] The square is also identified by Divorne and others as the basis of *bastides*' street-blocks.[37] An analysis of Grenade's layout reveals its 'hidden' squares (illus. 17). Using squares to generate geometrical built-forms cannot simply have been functional, however, for there were other ways to expand new towns which did not use squares or produce such rectilinear forms.[38] Moreover, and contrary to those who have argued for a

18 Layout of Mirande (France), *bastide* chartered 1281.

utilitarian purpose of these forms, a plan made up of a system of square-based units has a geometrical order that is far beyond the economic, pragmatic needs and requirements of royal administrators and local townsfolk. The rectilinear forms of the *bastides* were thus just as likely to have been chosen for their symbolic meanings as for the practical advantages they apparently gave.

Throughout medieval Europe, four-square, quartered new towns with rectilinear forms are thus present for different times and evident in different regions. Although they may differ in their details, the towns all show similarity through having as the basis of their forms a design based on the square. Sometimes the square is obvious in the town plan by having been systemmatically replicated throughout the urban landscape in the physical layout of street-blocks, as at Mirande and Geaune in Gascony,[39] or the new towns of Jaime II in Mallorca, as with Petra;[40] or by having been used to define the shape of the town's defences, as with Alfred's *burhs* (illus. 11, 18 and 19). Or sometimes the square is not immediately apparent. It can be manifest in more subtle ways. The square can be seen to make up the elongated shape of a marketplace, for example, even if it is not

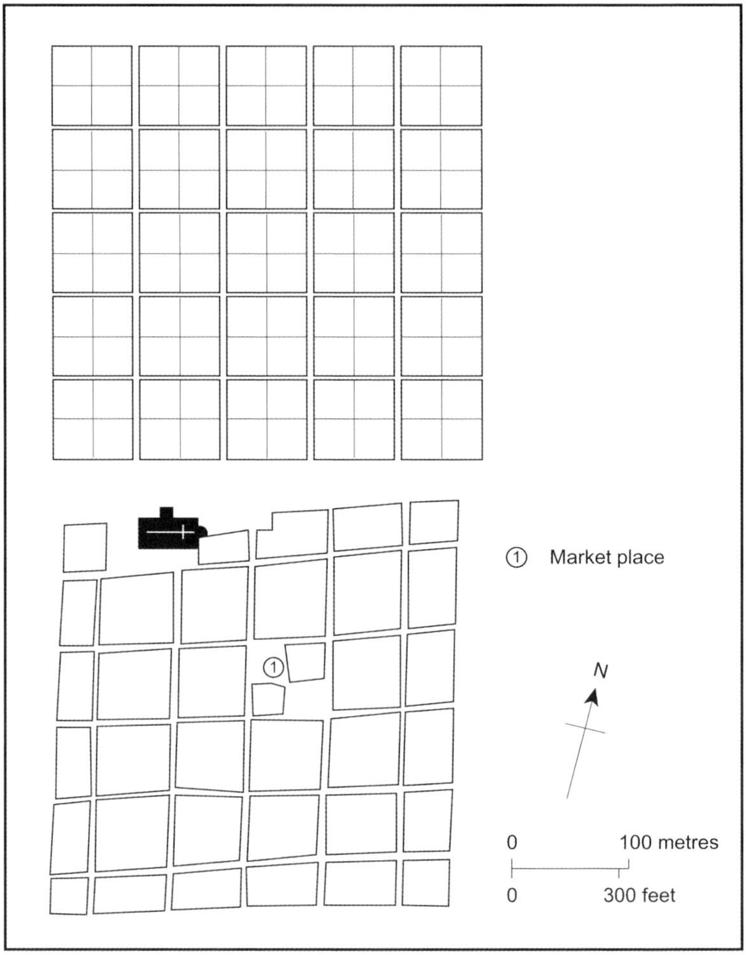

19 Layout of Petra (Mallorca), new town of *c.* 1300 with gridded layout (ideal plan and actual layout).

a 'true' square, or it might be 'hidden' in the quadrilateral forms of streets and plots – in their four-sided shapes, or in dividing a town into four quarters, however irregularly, by a crossed street-plan.[41]

Overall, then, it seems that the square, as an abstract geometrical figure, had certain qualities desired at the time these towns were created. In the rectilinear built forms of Alfred's *burhs*, the new towns of east-central Europe, and the *bastides* of southwest France, there was a common aim. Their squareness can be interpreted as a utilitarian response, as Randolph and others have outlined for the *bastides*. However, this would seem to ignore what square urban forms stood for at the time of their setting-out, a time which as we have seen was rich in cosmological symbolism based upon scripture; in particular, the geometrical model of the heavenly Jerusalem and its microcosmicism both as a history of the world and as a 'map' of divine order. In the cross formed by the grid-plan of these towns and in the axial arrangements of their streets, there is an overt Christian symbolism: the street axes quarter a town, as the wider world is quartered, the streets thus 'meeting and crossing themselves' to echo Lucian's allegorical description of Chester with its crossed streets. In differing but connected ways, therefore, a squared built form provided a link between heaven and earth: a sacred geometry that gave new towns a symbolic purpose.

Circular geometries and their symbolism

Compared with rectilinear built forms, examples of curvilinear form are more difficult to define. A place might have a circular-shaped outer defensive circuit, as was the case at Bruges with its city walls, for example. But what of internal urban layouts, the form of street blocks and house plots? Some have argued that the sinuosity associated with certain medieval urban landscapes was the product of complex circular geometries, but these views have received some criticism.[42] It seems that a purely circular built form – as opposed to curvature in general – was comparatively rare in the Middle Ages, and this in itself is surely significant. It could be explained by the practical advantages rectilinear forms had, but considering the particular significance attached to circular forms in Christian thinking – in imagining both city and cosmos – a paucity of such cases is all the more curious.[43]

If medieval new towns were being modelled on the celestial city, as seems to be the case, one possibility might be to do with how the heavenly Jerusalem was imagined differently during the course of the later Middle Ages at the time that many of the continent's 'new towns' were being formed. The rectilinear form widely used in new towns of the thirteenth century corresponds chronologically with a move towards showing Jerusalem more as it is described in Revelation, rather than as the circle of walls common earlier in the ninth and tenth centuries.[44] In this regard it is worth also noting an exercise in a late-twelfth century French *practica geometriae* that specifically describes how 'to infer the number of houses in a round city', but which comes from Gerbert's geometry, also of late ninth-century date.[45] This shift from circular to square *imagined* urban forms may thus have significance and explain why it is that circular *built* forms of new towns were so unusual in the twelfth and thirteenth centuries, the period in the Middle Ages when most new towns were created.

So while towns with varying degrees of curvature in their plans are to be found across medieval Europe, pure geometrically circular forms are generally uncommon. A number of towns with definite rounded built forms are present in Ireland. Prior to the main period of Viking settlement that took place in the tenth and eleventh centuries towns were being formed alongside monastic sites. These have attracted much interest in Ireland, not least because they confirm that urban life was not something imposed by outsiders on the Irish – by either the Norsemen or Anglo-Normans – but rather was present beforehand.[46] Although the form of these so-called 'monastic towns' is often more elliptical than truly circular, and sometimes based upon pre-existing structures, it is equally evident both at some of Ireland's most important ecclesiastical centres, such as Armagh and Kells, as well as more minor places.[47] In some cases, as at Armagh, this circularity is repeated to form a series of bounded concentric enclosures, in one of which sits the town proper, its marketplace positioned on the eastern side (illus. 20).[48] These enclosures are hierarchically ordered in terms of their degree of sanctity; the innermost, being most holy, contained the monastic church.

This hierarchical spatial ordering and the town's concentric circular-shaped enclosures appears to imitate the model of the heavenly city as depicted in ninth-century images where Christ as the Lamb of God is shown at the very core of and protected by a circular-shaped walled Jerusalem (illus. 38).[49] Contemporary written evidence supports this interpretation. A description preserved in the seventh-century *Synodus Hibernensis* states that

> There ought to be two or three *termini* [boundaries] around a holy place: the first in which we allow no one at all to enter except priests, because laymen do not come near it, nor women unless they are clerics; the second, into its streets the crowds of common people, not much given to wickedness, we allow to enter; the third, in which men who have been guilty of homicide, adulterers and prostitutes, with permission and according to custom, we do not prevent from going within. Whence they are called, the first *sanctissimus*, the second the *sanctior*, the third *sanctus*, bearing honour to their differences. [50]

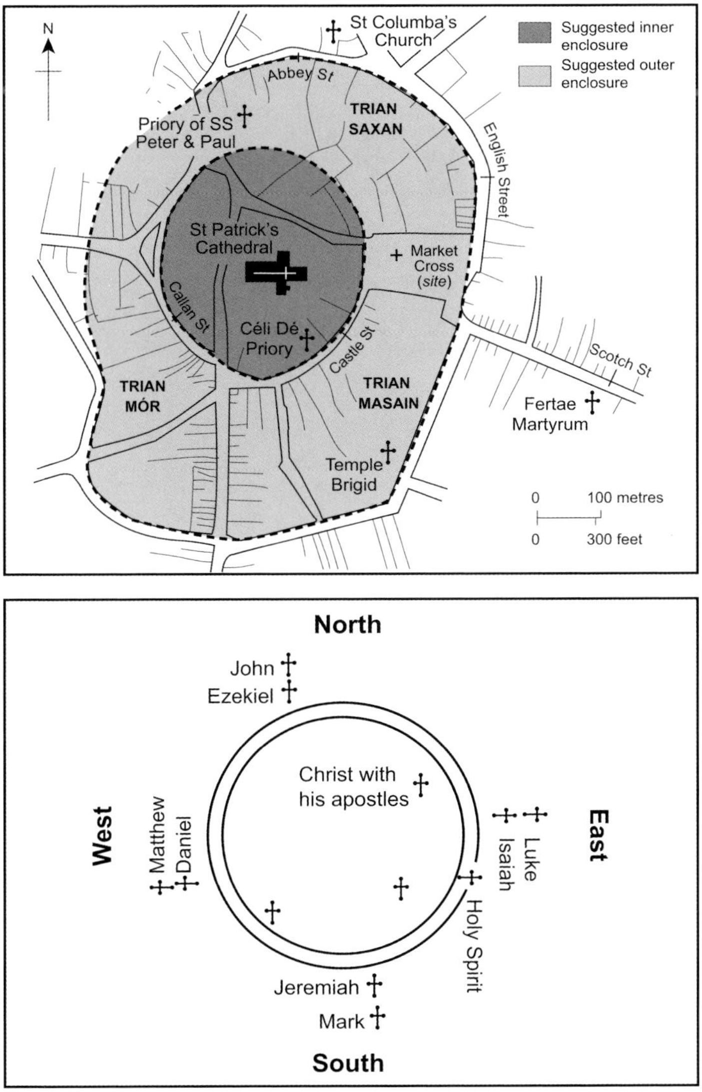

20 Armagh (Northern Ireland), curvilinear layout of monastic 'city' with its concentric enclosures (top), compared with the idealized 'monastic city' according to the Book of Mulling (Dublin, Trinity College Library, MS. 60) (below).

This concentric, circular form of the monastic 'city' also appears, diagrammatically, in the seventh-century Book of Mulling, in which two circles are shown with crosses, with names of the apostles and other saints marked inside and outside of them, the whole ensemble being orientated to the four cardinal points (illus. 20).[51] That these textual and visual descriptions of the imagined monastic 'city' correspond with the built forms of places such as Armagh is of significance in itself, but it is even more so if their sacred geometry (and sacral geography) is looked at more broadly, in the parallels between them and the circular form of contemporary depictions of the heavenly Jerusalem with its ordered and bounded sanctified spaces. In the built form of the monastic town in Ireland, then, is a model of the celestial city itself, replicated both in the outline of the enclosures within the monastic town – through their circular form – and in their function to create a spatial order that mirrors the divine order of the wider world, the innermost core – Jerusalem – being highest and closest to God. The circle here thus links the monastic 'city' of this earth with 'the Christian city of God'.[52]

The circular shape of Irish monastic towns may well be paralleled by others elsewhere in Europe where early Christian church-sites similarly form an urban focus.[53] There are circular built forms in Languedoc, for instance, but these were sometimes inherited from earlier rural settlements.[54] This is the case at Sarrant, for example, where a pre-existing village was rebuilt with circular-shaped outer defences encompassing an inner circular-shaped street, the focus of which is a church built on the site of an earlier chateau (illus. 21).[55] It seems that other cases were similar in this respect, such as Fourcès?[56] Until further study is undertaken, circular-shaped built forms do look to be rather absent from the later Middle Ages, yet instead, and perhaps significantly, there are many cases where square and circle geometries were combined in the physical forms of new towns.

Combining circle with square

New towns with rectilinear street-blocks, streets and plots and circular outer defences have a widespread occurrence across Europe, combining in a single design both square and circle. This geometric form appears on a unique manuscript plan of the town of Talamone in Tuscany, probably drawn up at the time the town was being created in the early 1300s (illus. 49). The plan drawing seems to have been produced for the purposes of planning the town and it shows a quadrilateral layout of street-blocks arranged in rectilinear form, regularly set out within a cross of streets, all contained within a circular-shaped circuit of outer defences.[57] The town's quartered quadrate plan is a close parallel to contemporary images of the earthly and Heavenly Jerusalem (for example, illus. 38 and 41).[58]

At Bristol too there is scope to compare the city's imagined and built forms. Although the fifteenth-century 'Bristol map' post-dates the town's foundation, the likelihood is that 'it is intended to

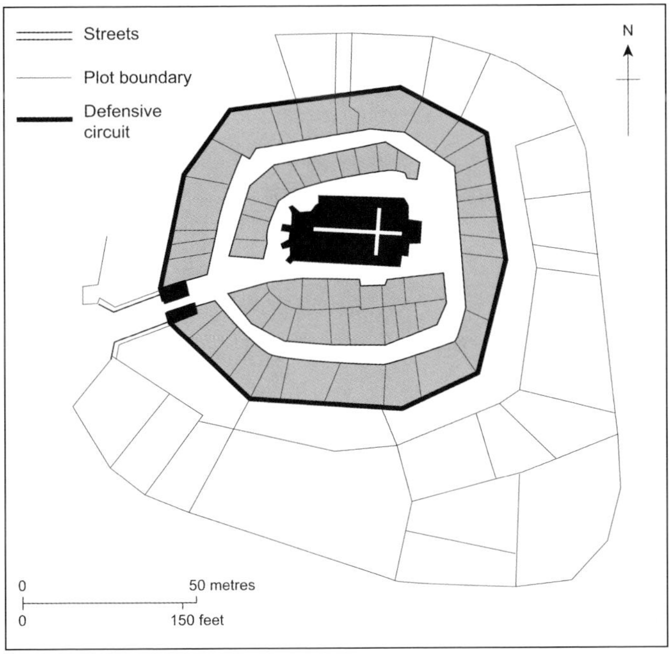

21 Sarrant (France), chartered in 1265.

represent the town towards the end of the Saxon period', in the eleventh century or before, when Bristol's urban core comprised a circuit of walls, roughly circular in form, and a cross of streets within (illus. 42).[59] The map appears at the beginning of a civic chronicle recounting Bristol's mythical foundation by a 'King Brennius', again suggesting it was a retrospective mapping of the city, visualizing the initial layout of Anglo-Saxon Bristol as a circle of walls and a cross-plan of streets and hence imitating the idealized geometrical form of the holy city (see chapter One). Here Bristol's circular and quartered geometrical form is similar in essence to that of Talamone. The fact that Talamone had this design in the early fourteenth century and Bristol in the tenth or eleventh centuries indicates the longevity this built form had during the Middle Ages. In both cases the built and imagined geometries point to an overall design deriving from a model based on the imagined form of the earthly and heavenly Jerusalem, itself depicted as a four-quartered (squared) circle (illus. 37 and 38).

Urban designs using circle and square geometries are evident in other medieval towns too. This is the case with the *bastide* towns of southwestern France, such as Créon, established in the early fourteenth century in Gironde by a seneschal for the English crown, as well as Geaune (Landes) of about the same date, and Monflanquin

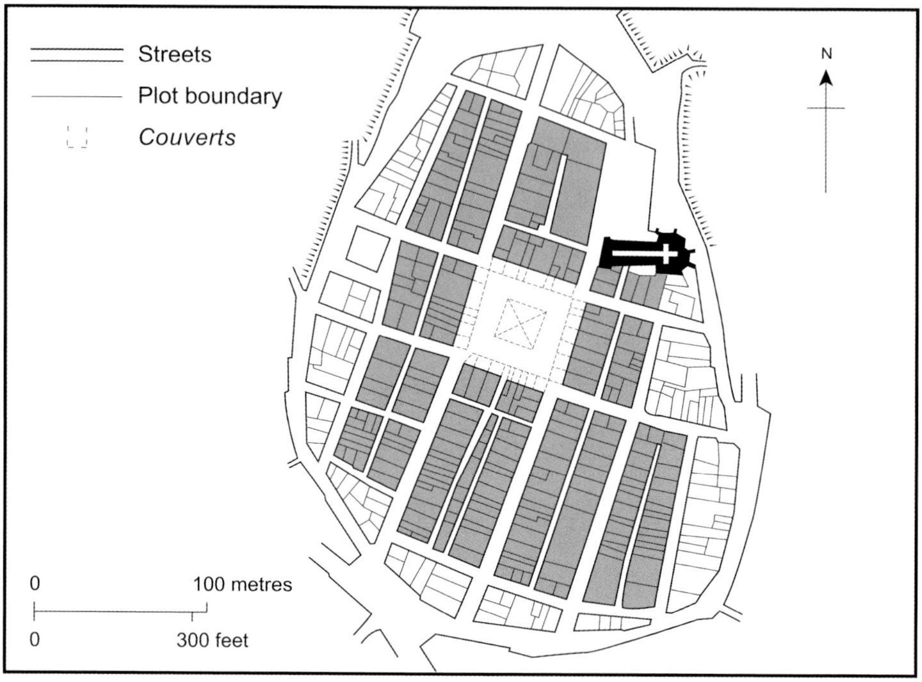
Streets
Plot boundary
Couverts
N
0
100 metres
0
300 feet

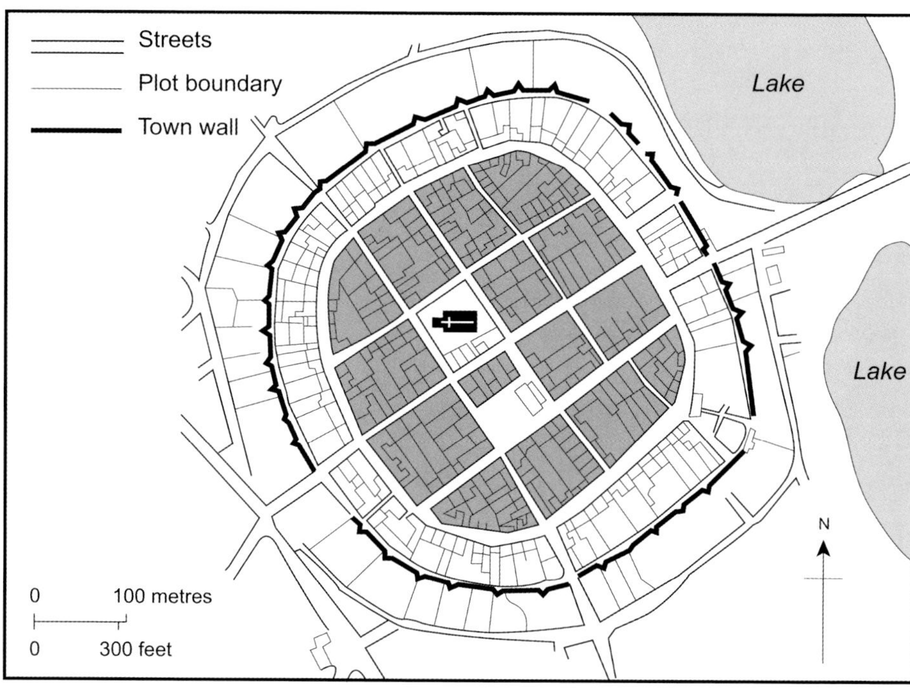
Streets
Plot boundary
Town wall
Lake
Lake
N
0
100 metres
0
300 feet

(Lot-et-Garonne).[60] The internal arrangement of these towns' street-blocks is rectilinear in form with a more or less circular defensive circuit placed around them. This is also a built form found in new towns of east-central Europe, such as Neubrandenburg (1248) and Strzelce Krajeńskie (1272–86), as well as in Italy, as with Cittadella (1220).[61] In each case the shape used to define the line of the defences of the town is the circle with the square being used to create internal order in its streets and plots (illus. 22). It is as if each shape had its own special symbolic purpose: the circle to protect the town, the square to order it.

A more sophisticated means of designing a town by combining circle and square forms is to be found in the plan of Terranuova, a new town created in the Florentine *contado* in the 1340s. Its layout of street-blocks and outer defences are rectangular but underlying them is a 'hidden' geometry of a circle divided into chords, each of which fixed the internal alignments of streets and plots, as Friedman demonstrates (illus. 23 and 24).[62] This approach used a circle in order to make a town quadrate, but it is only rendered visible by reconstructing the hidden geometry that created the town's design. Whatever way it was done, whether the circle and square were obvious elements of a town's plan (as with Talamone) or more subtle elements in its design (as with Terranuova), these two basic geometrical forms were combined to impose on the layout of the town the same spatial order that also structured the heavenly Jerusalem as well as the wider world, whose circles were likewise squared.[63]

Overall, then, medieval towns and cities with rectilinear and curvilinear built forms can be seen to be based upon particular use of square and circular geometries, or combinations of the two. These shapes are of course those that characterized imagined forms of both the earthly and heavenly city alike, as well as the wider Christian cosmos. The mirroring of these cosmological forms in the shared geometries of imagined and built urban forms points to a mutual, symbiotic symbolism, a common 'sacred geometry' connecting city and cosmos. This sacred geometry was materially inscribed into the medieval urban landscape. It is made visible to us today through studying the relict patterns and forms of medieval streets and plots from modern-day cartography and aerial photography. It is revealed through lines on the ground marked out as towns and cities were physically formed. Sometimes their geometries are rendered clearly visible by the material morphologies of urban landscapes, but there were also 'hidden' geometries used to generate the ground-plans of

22 Circle and square designs combined: Monflanquin (France), chartered 1256 (top), and Strzelce Krajeńskie (Poland), chartered 1272 (below).

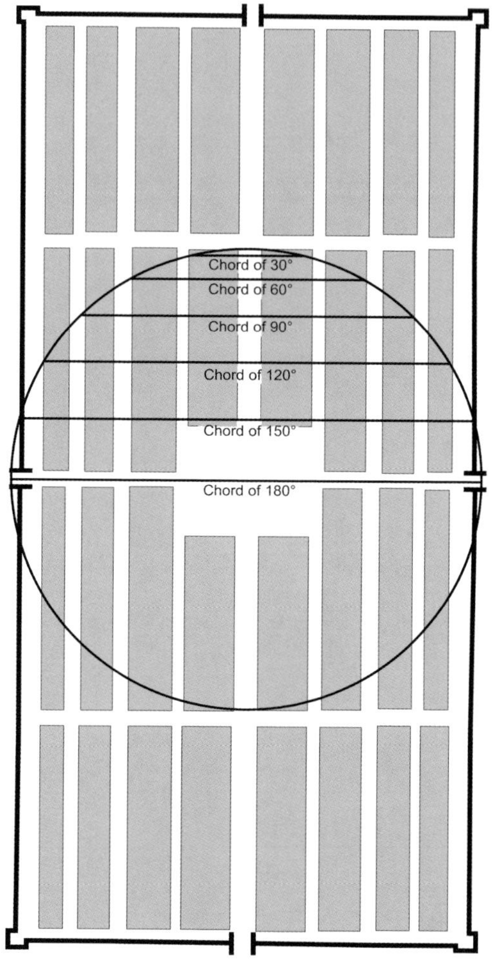

23 Circle and square designs combined: Terranuova (Italy), chords and plan, town founded 1337.

some new towns, imprinted in their built form. To consider further what these visible and 'invisible' geometries might have meant to those who produced them requires looking at how they inscribed on medieval urban landscapes a beauty which reflected cosmological order and aesthetic form.

AESTHETICS AND GEOMETRICAL BUILT FORMS

Geometry gave order to the medieval world. It was to be found in the cosmic 'body' and the human body alike, in the harmonic proportions of both, for 'the universe is a harmony of a multitude of orders'

24 Terranuova (Italy), view looking southeast across the new town through the remains of one of its gates.

(illus. 46).[64] This Timean cosmology was taught in the *quadrivium*, in geometry, alongside arithmetic, astronomy and music,[65] and repeated the Pythagorean proportional harmonies of the heavens, the 'music of the universe,' as Hugh of St Victor put it in his *Didascalicon*.[66] The numerical ratios of these heavenly harmonies, deriving from Ptolemy and Nicomachus, were set out and discussed by Boethius, as well as others afterwards, and thus had wide circulation in the Latin West.[67] Hence, proportion was a way of connecting the earthly and heavenly worlds, of providing a link between macro- and microcosmoi.[68] It was to be found everywhere in all created material and spiritual things, as Christian thinkers pointed out throughout the Middle Ages.[69]

Geometry, furthermore, gave beauty to these things, for it reflected God's eternal presence and reason.[70] This beauty derived from order, an order that God himself had created and which was made manifest in harmony and proportion.[71] As a part of this 'systematic theological ordering of the universe', Eco says, 'aesthetics became a philosophy of cosmic order' in Christian Europe.[72] These perfections found in the ordering of the universe were also being replicated on the ground by the creation of perfected urban forms, and the harmonies that kept the world in order also kept the physical layouts of towns and cities in order. What we can find in built forms of the Middle Ages is thus not only urban layouts being made perfectly geometrical in shape, but also a concern to arrange town plans proportionally. Their chosen underlying geometries were the same as those that ordered the world and gave it its beauty, a congruence that suggests that built forms had a cosmological symbolism more subtle than simply using circles and squares to make urban landscapes appear the same geometrical shape as the wider world.

Orthogonal forms

While proportion and ratio appear in built forms throughout the Middle Ages, certainly from the ninth century onwards, urban layouts have strictly geometrical, *orthogonal* forms only later on in the period. Why this was so seems to relate to a growing knowledge and theorizing of aesthetics and beauty in medieval culture based on an understanding of geometry, harmony and proportion in which geometrical perfection on earth was seen to emulate the perceived perfection of the universe, as was the case in art and architecture.[73]

A diachronic comparison of medieval built forms reveals a move towards geometrical perfection during and after the mid-twelfth century. Indeed, prior to *c.* 1150 it is difficult to find any examples of medieval towns and cities that have a strictly orthogonal form. The *burhs* of Alfred the Great, for example, while rectilinear and regular in their form, are by no means comprised of perfectly straight streets and plots set out at precise right angles to one another. Instead their forms contain many irregularities, as at Cricklade, where Haslam's plan of the town shows curvature in the lines of streets, particularly the east-west axis (illus. 11).[74] Similar irregularities are to be found in Alfred's other *burhs*, too, at Oxford, Wareham and Wallingford.[75] Even the outline of their defences is not perfectly square. Haslam regards the plan of Cricklade as 'a considerable feat of practical surveying', and suggests it is the work of 'a group of specialist surveyors attached probably to the king's court', and yet the town's layout is far from geometrically perfect.[76] The picture is very different a few centuries later.

By the end of the thirteenth century there were new towns to be found all over Europe with layouts that are highly regular in form, and perfect in their geometrical arrangements of streets and plots. These are towns of truly orthogonal form, with ruler-straight streets and street-blocks, and right-angled corners. It is as if whoever was laying out these urban landscapes literally straightened up their act. So in southwest France, the *bastides* of Monpazier, Grenade-sur-Garonne and Créon, mentioned earlier, all possess perfect orthogonal layouts, as indeed do others elsewhere in Europe, including the Florentine new towns such as Terranuova (illus. 14, 15, 17 and 23).[77] However, by no means do all the newly formed towns of this period possess an orthogonal layout. This is evident just by looking at those in east-central Europe, such as České Budějovice and Moravská Třebová, to name only two,[78] and likewise with the *bastides* of Gascony, where one town might have an orthogonal layout yet its contemporary near-neighbour does not.[79] This straightening up of town plans appears to begin at around the same time in various places across Latin Europe.

A move towards having straighter streets is discernible as early as the start of the twelfth century in new towns in northern Spain such as Puente la Reina, created in the 1120s on the Camino de Santiago, as well as in Normandy, as in the case of Breteuil-sur-Iton of about the same date.[80] Wholly orthogonal urban layouts only begin to appear later on during the twelfth century. One early example comes from the Welsh borders of England, where an extension in the built-up area of the small market town of Ludlow comprised an orthogonal design laid out in the 1150s or 1160s.[81] A century later and there are newly formed towns that have layouts that were likewise highly geometrical in shape, such as Villeneuve-sur-Lot (Agenais), with its plan evidently derived from a design based upon using right-angled triangles (illus. 25 and 26).[82]

Why it should be that some towns had such completely orthogonal plans while others at the same time were not given them (or only partially had them) requires some explanation.[83] A possibility is that it is to do with how geometry was understood by those who were involved in making the landscapes of medieval new towns, those

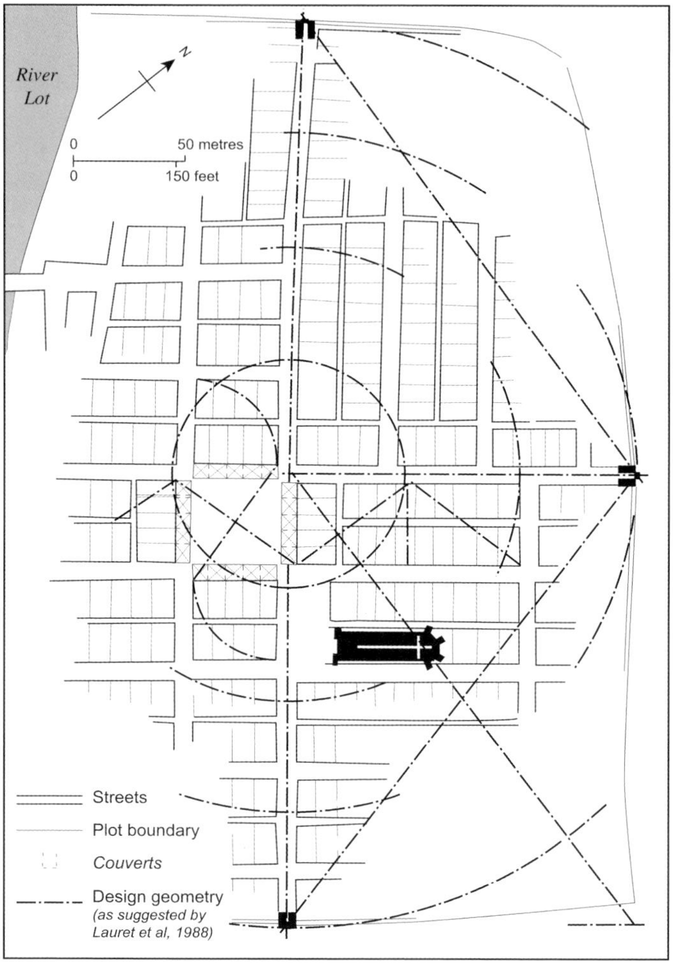

25 Geometrical composition and use of Pythagorean triangles in urban design: Villeneuve-sur-Lot (France), chartered 1260.

about whom so little is known. Could it be that geometrically perfect urban layouts were assuming greater importance in the later Middle Ages than they had done before? To resolve this requires us to consider the changes that were taking place in the use and meaning of geometry during the twelfth and thirteenth centuries, for the emergence of orthogonal urban forms so widely across Europe at one time suggests that it was a broad process at work, rather than a local or geographically specific one. It seems likely too, as we have seen, that such perfect geometries were not simply utilitarian in function – the effort to create them was surely much too great for that – but rather were used to satisfy some symbolic purpose. Such a symbolic purpose is suggested further by the presence of particular proportions and ratios

26 Porte de la Rue de Paris (fourteenth century), Villeneuve-sur-Lot (France).

in medieval urban forms, just as there were in architectural forms of ecclesiastical buildings built in this period.

Proportioned forms

The use of certain proportions in laying out urban landscapes is evident from analysing internal and overall physical dimensions in their plans. For example, each side of a rectilinear street block, or building

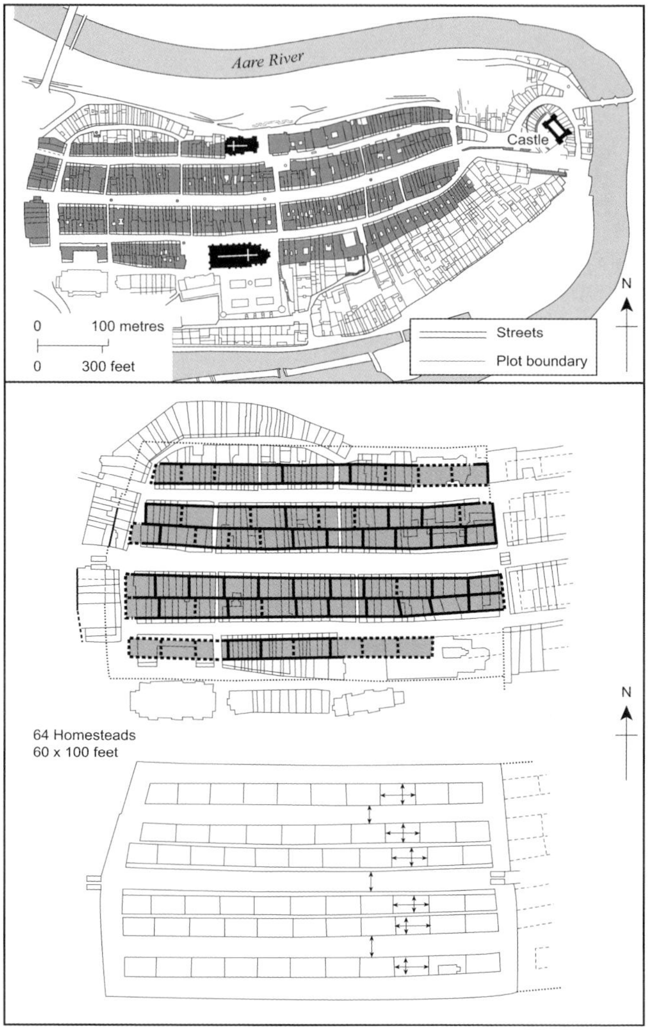

27 Zähringer new towns: Berne's town-plan (top) and plot dimensions (below).

plot, combines to give a length-breadth ratio, and these then define what proportions are evident in orthogonal and non-orthogonal plan forms alike. Take, for instance, the new towns created in the twelfth century by the Zähringer dynasty in the upper Rhein, an area where today Switzerland, Germany and France all converge.

While Hager describes the Zähringer 'new towns' as 'orthogonally laid out', examples such as Berne, Kenzingen and Breisach-am-Rhein reveal that their plans are simply rectilinear in form, and not orthogonal, since their streets and street-blocks are somewhat bowed rather than straight and perpendicular (illus. 27 and 28).[84] However, present nevertheless in these town plans are 'harmonic proportions', evident in the layout of their building plots, as in the case at Berne where they have a ratio of 3:5.[85] As well as in the dimensions of its house plots, this same ratio is present also in the overall arrangement of street-blocks in the western half of Berne, an area set out before 1191 (illus. 29).[86] Whoever came up with the design ratio used for the town's design clearly credited it with some importance. Giving Berne's layout an overall harmony thus had meaning. Somewhat simpler proportions are often to be found elsewhere, again indiscriminately between those new towns having or lacking orthogonality in their plans. In southern England, a group of new towns created along the Solent estuary in the later twelfth century had designs based on a ratio of 2:1.[87] A ratio of 2:1 was also behind the layouts of street-blocks in *bastides* in southwest France, formed by doubling the size of their square-shaped marketplaces, as is the case with Monpazier, Monflanquin and Miramont de Guyenne (illus. 15 and 22).[88] Friedman also points to the use of 1:2 proportions in the plans of the Florentine new towns created after *c.* 1330, at the unbuilt-town of Giglio Fiorentino, for example.[89]

The most curious and symbolically telling of all proportions used in town designs in the Middle Ages is based on root two (1:√2). At the moment it is difficult to say how geographically commonplace its use

28 Zähringer new towns: Breisach-am-Rhein's town plan (left) and plot dimensions (right).

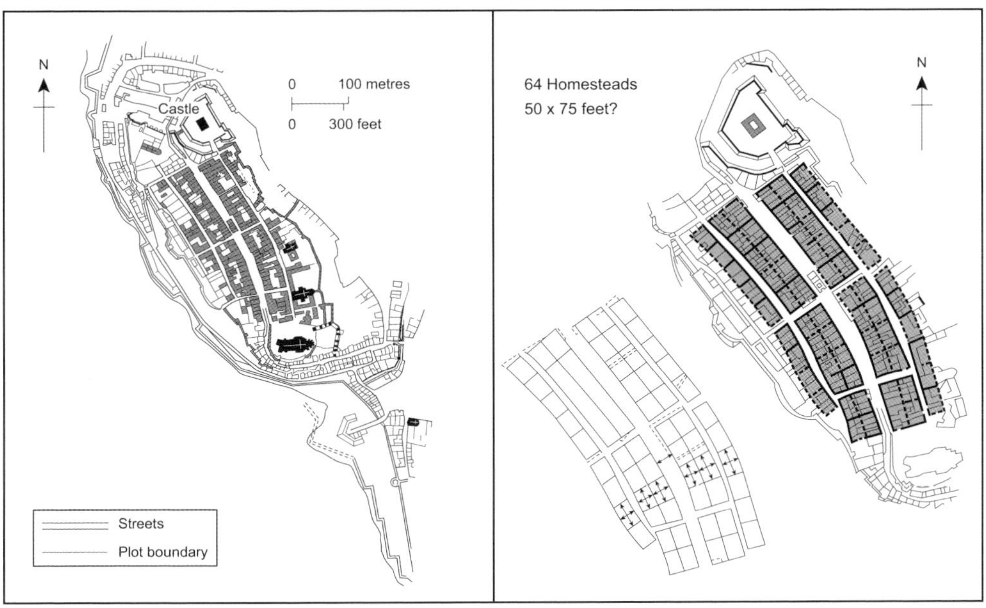

was, but it appears in town plans at opposing ends of the prime period of medieval new town formation. An early example is Bury St Edmunds, a town created in the mid- to late eleventh century just outside the gates of St Edmund's abbey, Suffolk, England, where in their analysis of its plan, Fernie and Gauthiez have proposed that 'the town was laid out geometrically and probably in conjunction with the abbey church since the axes of the church and town are the same', and because also both the town's street-blocks and the abbey's church are 'subdivided by the same geometrical proportion'; each based on a ratio of 1:√2.[90] Gauthiez notes too that root two was used at around the same time in the built forms of some new towns in Normandy, such as Lisieux and Le Petit Andely, and he compares these with the geometry of Bury's plan (illus. 30).[91] In all these early cases, the plans of the towns are far from completely orthogonal in layout, so the actual proportions are not always as neat as some interpretations would have us to believe. This is different with the example of Grenade-sur-Garonne, Eustache de Beaumarchais' new *bastide* of 1291, where the town is strictly orthogonal in form as well as designed using root-two geometry.

The proportioned layout of Grenade-sur-Garonne is especially sophisticated in not only having a plan derived from using the ratio 1:√2 but also in having this set out on the ground as a perfect geometrical form (illus. 17). Bucher first drew attention to the way the layout of Grenade was apparently derived at proportionally: he notes that the 'central square generates an east-west sequence' in which all the street-blocks are proportionally related to each other through being derived from the dimensions of their neighbours.[92] This idea is further discussed by Friedman who has pointed out that 'the geometry of Grenade is, of course, the system of rotation described by Villard de Honnecourt in his *Sketchbook*', also dating to the thirteenth century.[93] The generative geometrical design method of rotating squares was used in architecture in the later Middle Ages to create the pinnacles that adorned Gothic cathedrals.[94] In the case of Grenade this hidden geometry is made evident only by analysing the town's proportioned layout to reconstruct its underlying design. In this respect parallels emerge between the design use of geometry in creating both built forms and architectural forms in the Middle Ages; a shared sacred geometry and symbolic meaning.

Root two proportions of street-blocks in places such as Grenade point to an aesthetic purpose in the shapes of urban landscapes, and a complex use of geometry to generate their plans.[95] To have gone to

29 View of Berne looking west.

such lengths to create urban landscapes with such carefully proportioned forms, with their harmonic relationships between streets, would surely have required some expertise, and in the case of Grenade an ability not only to use the rotating square method (*ad quadratum*)

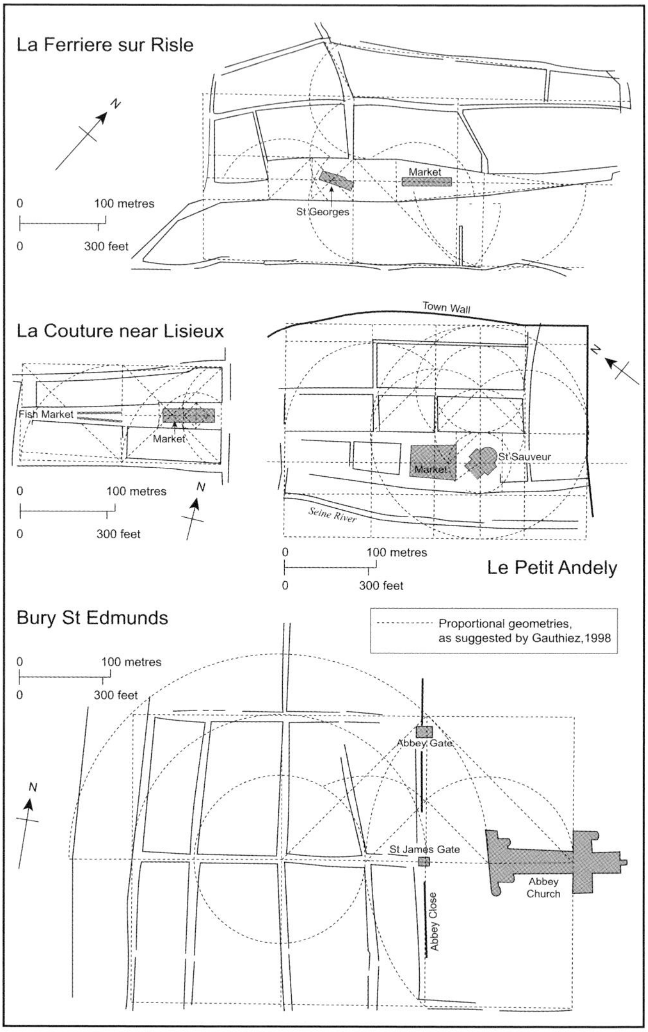

30 Use of 'root two' proportions to generate the ground-plans of Bury St Edmunds (England) and Norman towns in Normandy (France).

that was being used in architectural design, but also an ability to translate this abstract arrangement into a physical layout of streets and plots on the ground – surely no mean achievement.

Yet there is a proportioned layout that surpasses even that of Grenade for its geometrical ingenuity, used in the 1330s in creating Terranuova, a new town situated between Arezzo and Florence.[96] Here, as Friedman has revealed, the axes of the town's streets were based on a system of proportions derived from dividing a circle into chords, a 'sine geometry' that is seemingly unique to Florentine new towns and which was set out in Leonardo Fibonacci's *Practica*

Geometriae of the 1220s (illus. 23).[97] Terranuova's design again called not only for a sound understanding of an abstract theoretical geometry (and mathematics), but also a practical means of making the design a reality by setting it out accurately on the ground as a built form. Apart from suggesting that 'the grading of lots had an obvious economic logic' at the time, Friedman does not venture to suggest why these perfectly proportioned street-blocks were used by the Florentines and what they were meant to achieve.[98] Were they really just being pragmatic, or were these proportions and 'invisible' geometries being used for some symbolic purpose, to give their urban landscapes some particular meaning?

Making the 'invisible' visible

Terranuova's chords and their respective proportions conveyed harmony, while its orthogonal layout formed perfect symmetry. The presence of perfected and proportioned built forms may be understood as a reflection of a cosmologically derived aesthetics, an attempt to reflect in urban landscapes that same beauty to be found in the forms of other created things, including the macro- and microcosmoi. As Eco notes, 'the aesthetics of proportion was the medieval aesthetic *par excellence*' and by the eleventh and twelfth centuries, proportional harmonies were criteria used in discussions of beauty.[99] Later, by the thirteenth century, Aquinas and others were writing of God's perfection and looked to the natural world for examples of this, finding his beauty in what they saw: 'Beauty has to do with knowledge', wrote Aquinas in his *Summa*, 'and we call a thing beautiful when it pleases the eye of the beholder. This is why beauty is a matter of right proportion.'[100] 'The nature of the beautiful consists in general in a resplendence of form . . . in the duly ordered parts of material objects', wrote Albert the Great in *De Pulchro et Bono*, his work on 'beauty and goodness'.[101] Robert Grosseteste concurred: 'beauty is a harmony of proportions', he wrote.[102]

Beauty, then, was based on 'perfection' and 'due proportion' in material things.[103] One way in which proportional arrangements of material things were perfected was through their symmetry, for 'locating the centre of symmetry meant locating the way, the truth, the light'.[104] Of course, the thirteenth century was the time when urban landscapes apparently gained purer symmetry through having highly orthogonal forms with proportionally arranged plans, as noted earlier. This morphological change parallels these aesthetic theories on God's beauty, a change also paralleled in the ground-plans of Gothic cathedrals and churches of Europe that were likewise being straightened up at this time,[105] perhaps also to improve upon their 'order, disposition, eurhythmy, symmetry and beauty'.[106] Proportioned and geometrically perfect built forms thus fitted contemporary cosmological beliefs and medieval Christian thinking about the world's beauty and spatial order. First, the Christian (Neoplatonic) cosmos was ordered proportionally, especially the planetary spheres. Second, perfection was to be found particularly in those visible things that are harmonious and well proportioned, for these reflected God's 'invisible' beauty.[107] The basis of both was a theory of numbers and an understanding of the cosmos that stretched back via Boethius, Calcidius and Macrobius to Plato and Pythagoras.[108]

Since the numerical ratios built into the forms of new towns conformed to such universal proportions, the form of a proportionally arranged new town was microcosmically analogous to the proportionally arranged world, both having the same basis in geometry and embodying divine order.[109] Thus for Robert Grosseteste, writing in his *Praxis Geometriae* at the end of the thirteenth century, geometry, 'the science of lines, angles and figures serves to explain to us the very nature of physical things, of the universe as a whole, and of each part of it, of movement, of activity and passivity with regard to matter and with regard to the sense of sight and other senses.'[110] Presumably, too, a town's proportioned and geometrical form gained a greater sense of aesthetic beauty by having not only an invisible geometry in its design but in having a perfectly orthogonal form to render its geometry visible. This may explain the care taken in making some urban landscapes perfectly geometrical in form, and the apparent shift towards orthogonality apparent during the later 1100s onwards.

Orthogonal and proportioned layouts of urban landscapes revered God through achieving aesthetic perfection, but straightness in form also had symbolic meaning deriving directly from Holy Scripture. In likening Biblical exegesis to building 'that spiritual structure which is raised on high', for example, Hugh of St Victor in his *Didascalicon* had noted how 'the taut cord shows the path of the true faith'.[111] Straightness, it seems, was the path towards God. The straight path is in fact a metaphor used frequently in both the New and Old Testaments. In Isaiah, for example, we find 'Prepare the way of the Lord, make straight in the desert a highway for our God', and 'make the crooked paths straight', calls that are repeated in the gospels of

Matthew, Mark, Luke and John, while in Hebrews we are told to 'make straight paths for your feet'.[112] The streets of the new towns are these 'straight paths': their straightness pointed to God himself (illus. 15 and 25). In the case of Florence in the thirteenth century the government deliberately set out to make the crooked streets straight, for aesthetic reasons, to enhance the beauty and decorum of the city.[113] Of course, these passages from scripture were known well before urban landscapes were being straightened up during the twelfth and the thirteenth centuries. Their straightness in form thus came from an aesthetic desire derived from and based on scripture, as Hugh states in his reading.

Such geometrical perfection was likewise built into the newly emerging Gothic designs of abbeys and cathedrals of the later twelfth and thirteenth centuries.[114] Its presence in built form, through the use of proportions and symmetry, suggests that this cosmologically derived aesthetic philosophy crossed the boundary into the shaping of urban landscapes, that there was a common symbolism being used to shape new towns as there was to shape new Gothic buildings.[115] Perhaps this should not be surprising, as the two processes were occurring simultaneously during the Middle Ages. There are indications too that those involved in the shaping of urban landscapes were also builders, architects and masons, with shared beliefs about the meanings of forms. Witness the use of root two in the ground-plans of churches as well as in urban layouts like Grenade's. Straightness, then, like symmetry and proportions, were symbolic of God's beauty in the world, and geometrical arrangements of streets and plots symbolized cosmology just as it did in architectural form. This is what geometry meant at the time most new towns in Europe were formed in the twelfth and thirteenth centuries, a means of making the 'invisible' visible.

PART II

City-Cosmos Built

THREE

Founding a City, Founding a World

COSMOGONY, LORDSHIP AND URBAN FOUNDATION

'He was minded to found the one great city.'[1] In drawing this analogy between founding 'a city of stone and timber' and founding the world 'which our senses perceive', Philo of Alexandria is declaring that what God and the 'trained architect' had in common was a shared ability to create a 'city', a world.[2] It is an analogy that resonated throughout the Middle Ages, the Neoplatonic idea of the creator as *artifex principalis* being well known to Christian thinkers and depicted in medieval illuminated manuscripts showing God with dividers in hand, making the universe (illus. 47).[3]

This chapter examines how the geometrical formation of material layouts of towns and cities were connected with these conceptions of cosmogony, for just as geometry was instrumental in God's work in creating the world, so it was in the creation of new urban landscapes, in giving them their built form. As an act of creation, the role of the lord twinned with the use of geometry thus gave the whole process of town-founding a cosmogenic symbolism. This theme is explored first by looking at the social processes involved in founding a new town, and the analogies drawn between it and God's 'founding' of the world. The second part then explores how this analogy was further reinforced through the use of geometrical knowledge and techniques in the formation of urban landscapes, in coming up with designs for making towns as well as setting them out on the ground. This process of urban formation actually involved not just a political elite but a wide range of local agents, including surveyors and architects, whose use of geometry in the making of urban landscapes paralleled the same geometries used, it was believed, in the creation of the Christian world, in its cosmogony. In the minds of those involved therefore it was this practical yet symbolic geometry that made founding a new town analogous to, and a microcosm of, the supreme act of cosmogony – the foundation of the world.

The parallel between creating cities and creating the world actually had a firm Biblical basis. Significantly, Genesis itself recounts the foundation of a city when Enoch, Cain's son, 'builded a city, and called the name of the city after the name of his son'.[4] The foundation of other such cities is described too: 'the Lord scattered them [the tribes] abroad from thence upon the face of the earth, and they left off to build the city.'[5] In the New Testament, in Hebrews, a more mystical and metaphorical link is forged between city-building and the work of God: 'he looked for a city which hath foundations, whose builder and maker is God.'[6] The cosmogenic role of the heavenly and earthly city of Jerusalem, a metaphor for the Christian history of the world, is also told in Revelation, and through depictions of the holy city during the Middle Ages was given a geometric form (see chapter One). Symbolically, then, according to both Old and New Testament sources, city-building had a clear place and role in the making of the Christian world.

The analogy of the king as a creator of cities and God as creator of the world was also endorsed by the two principal foundations of medieval cosmogonenic thinking: Scripture and Neoplatonism.[7] For instance, Philo's analogy of the founding of the city, 'to satisfy the soaring ambition of some king', and the founding of 'the world which our senses perceive' had biblical and Platonic roots. Christian kingship and sovereign rule depended upon divine authority and order, as Gurevich reminds us:

> The highest position is occupied by the monarch appointed by God. Just as the world (the macrocosm) is directed by God, and the human body (the microcosm) by the soul, so is the body politic directed by the monarch, whose relation with his subjects might be compared to the relation between the head and the limbs. The power of the monarch does not depend on the will of the subjects. The monarch is subject to one God alone, whom alone he serves (*rex* – minister Dei).[8]

It is a microcosmic view of social order, reinforced by ideas from Timean Neoplatonism (for example in Macrobius, Calcidius and

Boethius).[9] This idea that God and king were connected had wide circulation in the Latin West. John of Salisbury, for example, wrote in *Policraticus* that 'every ruler is "a form of earthly image of divine majesty"'.[10] The analogy between king and demiurge is evident in other twelfth-century texts too, such as the *Plaint of Nature*, in which Alan of Lille drew upon ideas from Neoplatonic microcosmicism as well as Genesis when he asserted, 'for in heaven, as at the pinnacle of an earthly state, the eternal commander has his imperial abode; from all eternity the order had gone out from him that each and every thing should be inscribed and made known in the book of his Providence.'[11] In his sermon deriving from Matthew 21:2, Alan draws an even closer parallel between God, kingship, cosmogony and the city: 'the King of heaven and earth, who commands the winds and the sea, created this world in the image of a city.'[12]

All this was known – not least in cloister and court – at the same time that many new towns were being founded in Europe by sovereign lords. In the case of England, for example, Alfred of Wessex himself, as well as founding the *burhs* that saved him his kingship, translated Boethius' *Consolation of Philosophy*, while John of Salisbury was 'well acquainted with important figures in twelfth-century secular life, especially the young Henry II' whose reign saw a steep rise in the number of new towns founded in England.[13] It seems likely therefore that those who were overseeing the founding of new towns had an awareness of the cosmogenic symbolism that this process invoked.

The symbolic role of kingship in founding a new city is evident not only in political philosophy drawn in part from Calcidian Neoplatonism but also, later in the Middle Ages, from Aristotle's *Politics*. Aquinas' *Commentary on the Politics* did not extend as far as Book VII and so it excludes Aristotle's discussion of appropriate forms of city layout for 'safety and good appearance'.[14] But in *De Regimine Principum*, Aquinas attends in some detail to the matter of how to found a city, and observes how 'the foundation of a city or kingdom can be suitably derived from the form of the foundation of the world.'[15] Dealing with its site, layout and its content, his discussion in Book I of *De Regimine* follows the *Politics*' advice quite closely, combining both Platonic and Aristotelian microcosmicisms:

> The one who founds a city or a kingdom cannot produce human beings and places for them to inhabit and other things to support their lives all over again, but must use those things preexisting in nature, just as the other arts take the material

> of their operation from nature. So, for example, in the practice of their art the metal worker uses iron or a builder wood and stone. Therefore, first of all it is necessary for the one who founds a city or kingdom to choose a suitable place that will preserve the inhabitants by its healthfulness, will suffice for their sustenance by its fruitfulness, will delight them by its pleasantness, and will render them safe by protecting them. If it should lack any of these advantages, the place will be more suitable to the extent that it has more of the necessities that I mentioned or the more important of them.
>
> Then, it is necessary that the one who founds a city or kingdom distinguish the parts of the chosen place according to the exigencies of those things which the perfection of the city or kingdom requires. So that if a kingdom is to be founded, it is necessary to decide what places are suitable for constituting cities, what for villages, what for castles, where the studies of letters should be constituted, where the knights should exercise, where business should be conducted, and so on for all the things that the perfection of the kingdom requires. But if the work of foundation is for a city, it is necessary that the one who founds it decide what place should be appointed for sacred matters, what for handing down law, what for individual artisans. Further, it is necessary to gather together the persons who are assigned to suitable places according to their duties. Finally, one must provide for necessities to be at hand for individuals according to their individual constitutions and states; otherwise, the city or kingdom could never last.
>
> I give only a summary of those things which pertain to the office of king in the foundation of a city or a kingdom, and I have chosen them for their similarity to those in the foundation of the world.[16]

This guidance deals with Aristotle's concerns over founding a new city,[17] but it also goes further by justifying its creation as an analogy, the model being provided by 'the foundation of the world', itself a motif with obvious Platonic resonances of God 'the wise architect' (*sapiens artifex*), using his compass and square to fashion the world like a city.[18] The Neoplatonisms do not stop there, either. Aquinas' thoughts on art and beauty were based on principles that connected God with geometrical forms, especially forms with order and proportion, themselves familiar to Neoplatonist Christian thinkers of earlier times (see chapter Two).

The perfection of form for Aquinas was a reflection of God's beauty and presence in the world. In part this was an aesthetic derived from Augustine and Boethius, and their Pythagorean concepts of hierarchy and harmony, blended of course with Christian doctrine from scripture, and the idea of the need to 'keep to the straight and right path' towards God.[19] The geometry of Aquinas' aesthetics is thus a Platonic one, but one that did not stand in opposition to the aesthetics of urban form that readers of Book VII of the *Politics* would have found. Aquinas' broadly Neoplatonic aesthetics of form complemented Aristotle's conception of an urban aesthetic of orthogonal order. Here again, then, there was in the city a broader 'world', revealed by shared forms with their proportioned and harmonious relationships to one another – just as in the macro- and microcosm that was the world and man.

Through commentaries on Aristotle's *Politics* advice was thus gained from 'the Philosopher' about how best to build a new city that 'will delight them by its pleasantness' and, at the same time, 'render them safe by protecting them'.[20] In both respects, through the aesthetics of its form and through its 'lofty' and secure position, Aquinas' 'city' was made to be closer to God. An example of this is Ptolemy of Lucca's development of Aquinas' *De Regimine Principum*, a continuation of the text begun from earlier work carried out by Thomas Aquinas and his commentary on Aristotelian political thought but with Ptolemy's own experiences of government in the city republics of northern Italy, with which he was familiar.[21]

Like Aquinas, Ptolemy sets out how 'it is necessary to establish a city' and 'how many parts are in it', as well as the idea that 'all lordship comes from God, as from that First One who exercises lordship', doing so by citing scriptural sources as well as Classical ones, particularly Aristotle's *Politics* and *Metaphysics*.[22] Although principally Aristotelian in perspective, his views were perpetuating a microcosmicism present in earlier Neoplatonic work,[23] about how the earthly city and the 'city' of God were both seen as being the work of sovereign lords, one in the image of the other. Ptolemy's rhetoric was to some degree influenced also by his wishing to justify government in Italy by city republics, which included his own city of Lucca.[24] These city-states were themselves founding new towns at this time, such as Terranuova and Giglio Fiorentino by the Florentines, and Cittadella by Padua, each with their geometrical built forms (illus. 23). As far as the government in Florence was concerned, 'new-town committees' were set up to oversee this process, run by men with varying degrees of autonomy as part of their 'civic obligation'.[25] Guido Vernani's early fourteenth-century commentary on the *Politics* likewise had a specific non-intellectual audience, the mercantile elites and podestàs of Italian city-states.[26] In essence, then, while Ptolemy and others around him of like mind might have been disparaging of kingship (he regarded it as despotic), the political animals that were the city republics nevertheless acted as kings had in the foundation of their new towns. They too were akin to the Neoplatonic model of the senate, as head of the state, issuing its demands to the lower orders for them to be carried out. Therefore it seems certain similarities existed in the way that new towns were being founded by sovereign authorities in Europe during the Middle Ages.

Whether initiated by kings or city-republics, bishops or princes, the process of founding towns was mirroring a cosmogenic model based upon Holy Scripture and moulded by ideas from the *Republic* or *Politics*. Furthermore, this model, since it was based upon the authority of God and his founding of the world, also legitimized not only the social and political positions of these founders but also their symbolic role in town founding. Moreover, the *De Regimine* was written with a specific audience in mind, and in this context, as a 'mirror of princes', it is not too much to expect it to have been read by those within the royal court.[27]

The role of the lord

The genesis of a new town generally took a number of years before it was actually finally complete. This is evident with the new towns founded in England and Wales by King Edward I at the end of the thirteenth century. In the case of the founding of Winchelsea in Sussex on the southern coast of England, the king first gave directions to find a site suitable for the new town in 1280 but it was only in July 1288 that the town was complete in having street-blocks ready set out and building plots provided for possession.[28] At Flint and Rhuddlan in north Wales, sites for the towns were initially inspected by the king's clerks in summer 1277 but the towns were not actually settled by burgesses until later in the following year, with charters of borough privileges formally granted by the king.[29] This lengthy and hierarchical process of decision-making began with instructions issued by the

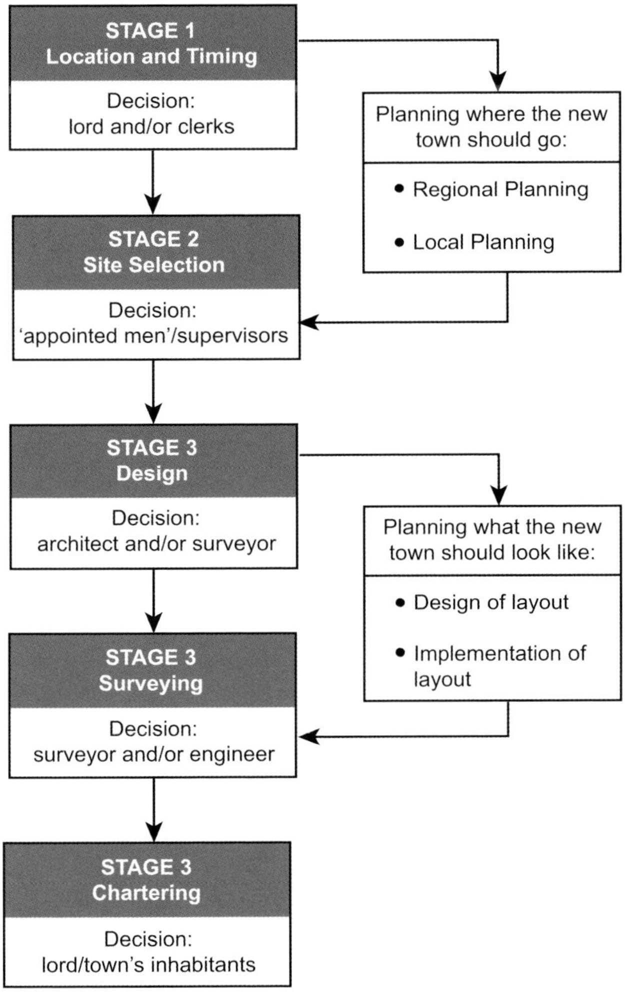

31 Stages in the formation of a medieval new town.

sovereign lord under whose authority a town was to be created (illus. 31).[30] These were then followed by decisions of other individuals who were called upon because they had experience or expertise in, say, the matter of negotiating between local parties over access to land for the site of the town or deciding upon its privileges perhaps, as was the case with some *bastides*,[31] or else they were skilled in the design and laying out of a town on the ground. Only after all this was a new town finally founded legally, being granted a charter to confirm its status and privileges, itself an auspicious occasion.[32]

In founding a new town the highest authority was the sovereign overlord – the king, prince or duke – but new towns were established also by a wide variety of lesser lords, both secular and ecclesiastical.[33] Founding a new town was at the discretion of the sovereign lord, for whoever the subject lord was, their lands and their privileges were held by consent, and if they fell out of favour with their overlord then what had been granted to them could just as easily be taken away.[34] In some cases a sovereign lord issued instructions for towns to be founded on his behalf, instructions that were carried out by subordinates of the royal household, as with seneschals in Gascony serving the English and French crowns, the *locatores* in east-central Europe employed by dynastic lords like Duke Barnim of Pomerania, or the men appointed by Edward I to 'plan and assess' the new town of Winchelsea.[35] New towns were also being founded throughout Europe on ecclesiastical lands by episcopal and monastic lords. Soon after victory over Harold in 1066, William the Conqueror established an abbey at Battle, and shortly afterwards a new town was founded by the abbey outside its gates.[36] In the twelfth century Archbishop Wichmann of Magdeburg populated lands between the Elbe and Oder rivers in Poland and eastern Germany with new towns, providing them with laws that favoured the settlers drawn to them.[37]

So founding a town – whomsoever it belonged to – was a process in which the lord's role was primary, and in founding towns upon their lands these lords shared a common purpose. Socially and politically, they stood at 'the pinnacle of an earthly state' (as Alan of Lille had put it), overseeing the process of urban formation. There is therefore a parallelism in the creative role played by these lords and *the* Lord, in which a city, like the world, is brought into being. In founding towns, then, these lords were acting as God had in creating the world. From a Neoplatonic perspective, in which the 'state' on earth mirrors the 'celestial state', they each were the centre of reason, the 'citadel' from which orders were given out, on earth as in heaven.[38] But the founders of the towns not only imitated God's role in Creation, acting as God had according to both Genesis and the *Timaeus*, issuing commands to the 'lower orders', and creating on earth a microcosm of the celestial archetype: they were also themselves part of the divine order which linked them with the Creator himself in the chain of cosmic being. The fact that such lords derived their literacy and learning from scripture and Christian Neoplatonism makes it all the more likely that they were in fact aware of these cosmogenic overtones in their symbolic role as town founders.

Assuming, then, that some sovereign authority, say a king, had instructed that a town is to be founded, and that the choice of site and other such preliminary tasks had been begun by appointed individuals or committees working on behalf of their superiors, how then did the formation of the new town proceed, and in what ways was this also symbolically cosmogenic? Written records made at the time are not especially forthcoming on who was doing what, particularly so for those stages in the process lower down the chain of command which were concerned with the designing and surveying of the new town (illus. 31). What little evidence there is tells us something about how urban landscapes were physically made, and in so doing, once again it will be geometry that provides the key to linking city and cosmos, and more specifically urban formation and cosmogony. Geometry, it will be seen, was used by those who created urban landscapes in the Middle Ages, both in the design of a town and in laying it out. To see this means looking in detail at the people who were involved in this part of the decision-making process, a range of individuals who were trained in the art and science of geometry.

GEOMETRY AND URBAN FORMATION

Images of God the geometer, *artifex principalis* complete with compass in hand, were contemporary with the time when so many of Europe's new towns were being brought into being, between the ninth and fourteenth centuries.[39] The microcosmic meaning and Neoplatonic origin of these images was commented upon earlier in chapter One. Here, what is important is that they reveal the symbolic connection drawn between God, who had used geometry to fashion the universe, and those on earth who were using geometry to create their towns and cities, thus linking them cosmogenically.

There are two key issues to address in this respect. The first is finding out who these individuals were, and the second is how – like God's work – theirs was based on the use of geometry. The evidence for both comes from a close reading of the built forms of urban landscapes alongside contemporary written sources. Both tell us about the work that was done to create new towns – about the knowledge and skills that were needed to design and survey urban landscapes – and it is this that points towards a cosmogenic symbolism of urban formation, since those who were expert in geometrical knowledge and skills were the ones doing the work. Their activities were based on practical and theoretical geometry which was itself derived from contemporary cosmographical discourse through a tradition of geometrical work that stretched back in time to antiquity via the likes of Gerbert, Isidore, Boethius, Macrobius and Calcidius, as well as Roman surveyors' manuals that survived into the Middle Ages as various texts known as the *corpus agrimensores*.[40] Cosmogenic symbolism thus not only existed in the practices of individuals, in the use of geometry to design and survey new towns, but also through being part of this long tradition which linked together geometry and cosmogony.

Who, then, did a sovereign lord instruct to work on forming a new urban landscape? As with so many aspects of life in the Middle Ages, there are no surviving accounts from the time that tell us what those involved in making urban landscapes thought about what they were doing. Indeed, it is rare even to find such individuals named in contemporary sources and for their activities to be recorded at all, except occasionally in the construction of castles or religious buildings by master architects and masons employed to oversee the work.[41] This situation obviously poses problems if we wish to tease out the cosmogenic symbolism of medieval urban formation. Surviving descriptions of medieval cities which comment on aspects of their physical forms are post hoc accounts: none were contemporaneous with the time of design and planning that first brought them into being.[42]

Architects, urban design and geometrical knowledge

In writing on the building of Europe's cathedrals, Gimpel comments in passing that 'life's vagaries led some of these architects to become town planners and thus they designed new towns'.[43] Whether or not medieval architects and master masons worked on new town projects when less prestigious or remunerative jobs were to be had is hard to say, but certainly they would have offered town founders relevant expertise, based upon their geometrical knowledge and skills, in either drawing up designs or setting out ground-plans, or both. There is indeed some evidence that points to this process.

There are numerous illustrations from the four centuries after 1000 that show either a king or a bishop instructing their masons and architects in building work and construction (illus. 9 and 48).[44] Architectural elevational drawings of buildings also exist from the thirteenth century onwards, showing that designs were sometimes worked out first on parchment and then applied in construction work (illus. 32 and 33).

32 Elevation drawing for Strasbourg Cathedral, *c.* 1250–60 (black ink on two pieces of parchment).

As surviving examples show, this was the case for cathedral building, as well as for some mercantile buildings in towns.[45] Geometry was used to compose these drawings, as Villard de Honnecourt makes clear in his *Sketchbook* when he notes 'here begins the art of the lines of drawing as taught by the discipline of geometry which facilitates work.'[46] So too does Roger Bacon in his *Communia Mathematica* of the later thirteenth century, who remarks on the various uses of geometry; one

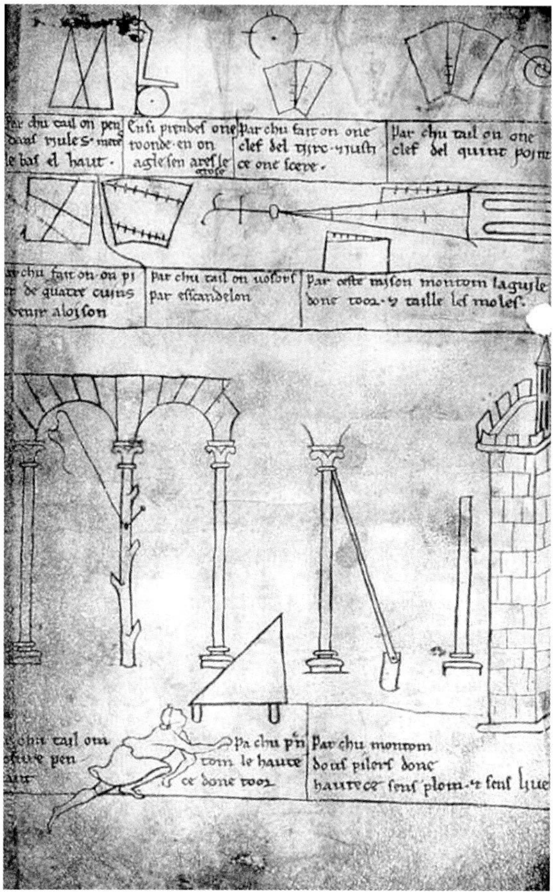

33 'In this manner one measures the height of a tower', Villard de Honnecourt, *Sketchbook*, *c.* 1230.

being 'for the use of men in governing families and cities', which included, significantly, 'how to draw cities, castles, houses and so forth'.[47] In the anonymously written *Artis Cuiuslibet Consummatio*, a book of practical geometry written in northern France around 1200, the author shows with the use of a diagram how to calculate 'the number of houses to be obtained from a known surface', and how 'to infer the number of houses in a round city', two exercises that would have been helpful in the design of new towns.[48]

Not only was geometry used to give form to drawings and plans, therefore, but it was used also in the process of their formation, through the act of drawing and calculating. Bacon's comment suggests moreover that there were those using geometry to draw cities, presumably in order to design them and not simply to represent them in pictorial form.[49] In which case, it seems highly likely that architects were the individuals most equipped for doing this drawing and designing. In contrast to the drawings made of buildings, however, surviving *drawn* designs for new towns or urban landscapes are conspicuous by their absence.

Most urban mapping was undertaken after the fact, except for one very instructive example that shows the new town of Talamone in Tuscany, apparently at the time of its making (illus. 49).[50] Similar such drawings may simply have not survived, or perhaps they were made but destroyed through being sketched out on a clay or wax tablet rather than on more expensive parchment.[51] In other cases the design may have resided in the architect's own mind and only taken material form during the physical laying out of the town. For example, in setting out new town defences around Ardres in Flanders, a Master Simon, 'the master of the geometrical work, proceed[ed] hither and yon with his stick in a magisterial way, as he measured not so much with his stick as the ruler of his eyes the progress of the work he had conceived in his mind.'[52] Likewise, somewhat analogously, Thomas Aquinas had noted 'a house, for instance, is true if it turns out like the plan in the architect's mind.'[53]

There are further signs that architects were forming new towns in the Middle Ages and using geometry to do so. In the case of the Florentine towns studied by Friedman, the appointed committees overseeing town foundation employed skilled architects, as contemporary accounts make clear. The members of the committees had themselves 'no professional training as builders, surveyors or designers', and instead they secured an 'architectural advisor' who not only dealt with 'technical issues', but also more administrative ones as well.[54] One of these men was Bachino Cambiuzzi, described at the time as 'chief master and overseer of the works', the person to whom the committee had delegated the job of town design and building.[55] On the matter of design per se mason-architects such as Giovanni di Lapo Ghini and Neri Fioraventi worked on aspects of cathedral building as well as new towns such as the design for the intended but never built town of Giglio Fiorentino with its 2:1 proportions (see chapter Two).[56]

While the better than average written sources of the fourteenth-century Florentine towns 'record the presence of architectural professionals at the town building sites, as advisors, supervisors, administrators, official representative of government, and planners',[57] for other new towns, and especially for those of earlier periods, contemporary written accounts are not so forthcoming about these close connections between architects and urban design. Instead, the forms of the towns themselves – the designs on the ground considered in the previous chapter – provide the evidence and can be used to suggest that architects were sometimes responsible for them. Perhaps the clearest example of this comes from those new towns laid out with plans that are based on root-two geometry mentioned before, such as Grenade-sur-Garonne and Bury St Edmunds (illus. 17 and 30). With Bury the co-presence of root two in the layout of the town's streets as well as in the ground-plan of the abbey church – both contemporary with one and other – suggests that an architect was behind the overall design, spatially uniting church and town through their shared geometries.[58] With Grenade the ratio of $1:\sqrt{2}$ is used so precisely to set out the town's street-blocks that the person responsible surely had a very good idea of both how to create the design and then set it out accurately on the ground, and since root two was primarily used by architects in the building of cathedrals and churches, in making pinnacles in particular, the finger of suspicion for the layout of Grenade points firmly in the direction of an architect, albeit, as is usual for this period, an unnamed one.[59] Perhaps in these cases, as contemporary images of architects working on buildings suggest, and as evidence from the Florentine new towns shows, an architect was commissioned to carry out the work on instruction, though of course how much consultation took place between the architect and his employer(s) is open to debate. In urban and architectural design, geometrical knowledge was applicable to the theory and practice and both.

Mensores and surveying the town

As well as architects there were others working on creating new towns using geometry. One group was the 'measurers'.[60] They too had the necessary expertise to lay out on the ground whatever features made up a new urban landscape. They are thinly documented, but by the twelfth century *mensuratores* are referred to in civic records in Italian cities such as Pisa and Milan, where they worked for the city's government measuring existing lands, particularly in property disputes.[61]

In Flanders, too, there were measurers working. They are recorded in the thirteenth century in Douai and Bruges, for example, as well as Ypres and Mons.[62] At Bruges, measuring for the city was done by John *lantemetere* in 1282, then again in 1284 when he is instead called *geometricus*, and again in 1285, while in 1287 others like him were employed to work on the city's new defences, both during the course of their construction and afterwards as inspectors of the works.[63] The measurers were therefore used by cities in different capacities, known by a variety of terms (*geometricus*, *agrimessor*, *mesureur*, *mensurator*, *landmeter*), and in some cases they were architects or masons employed as 'master of works' (*souvrain des ouvrages*), as at Mons where Jean le Roy is named in civic records in 1316 attending to works in the town.[64] It seems, then, that 'measurers' were simply those individuals who were skilled in measuring, and given the task to measure either what was there already or to set out new works (illus. 34).[65] That they were skilled in 'geometry' is suggested not just by the designation *geometricus* given to them, but by contemporary Flemish accounts such as Lambert of Ardres' description of Master Simon who, in laying out new earthwork defences at Ardres, is described by him as 'the master of the geometrical work'.[66]

While it seems likely that these measurers were employed to work on the layouts of new towns, actual examples are difficult to find. With the founding of the *bastide* of Baa near Bordeaux in March 1287, a 'master' measurer does appear to be involved with physically laying out new streets and building plots, a Master Gérard de la Tour, '*assignato ad ordinandum bastidam*', who is mentioned as '*magister operum nove bastide*' in the exchequer accounts of the town's founder, Edward I.[67] At the same time a large cord had been purchased for two shillings, presumably for the purpose of surveying, and later in the year there is payment made for expenses including measuring the streets of the town, a matter handled by the king's administrator

34 Measuring the city with a cord: ceiling painting, chapel of St Clemens, Schwarzrheindorf, Bonn (Germany).

on site, Richard de Escham.[68] Such measurers, or surveyors, of new towns are shadowy individuals too often obscured by the names of those for whom they were working.[69]

Setting out a new urban landscape thus involved measurers expert in the use of geometry. Whether these individuals created urban designs themselves, or simply followed through designs handed to them, is not made clear in written accounts. It perhaps depended upon who they were, and their individual way of working. What is clear is that different types of measurer were working in the twelfth and thirteenth centuries, distinguished by their knowledge and use of geometry (illus. 33). This differentiation between the users of geometry is revealed by *practica geometriae* texts written in the thirteenth century. For example, in his *Practica Geometriae* of *c.* 1220 Leonardo Fibonacci distinguishes between audiences for his book, between 'those who would work following geometric demonstrations and those who would proceed following common usage or, as it were, lay custom'.[70] As Victor points out, those who measured according to 'lay custom' were the surveyors who calculated their measurements using ropes and other physical means, while on the other hand 'geometric demonstrations' were based on theoretical geometry, derived for instance from Euclid's *Elements*.[71] Fibonacci's two audiences may not have so neatly divided in reality, however.

One of the conclusions drawn by Friedman in his analysis of the built forms of Florentine new towns is that Fibonacci's chord tables were the probable source used to devise and set out the proportioned plan of Terranuova (illus. 23).[72] If so, not only were surveyors using these texts to do their measuring work, and making use of theoretically derived geometries, but they were doing so in order to create new urban landscapes. Other *practica geometriae* texts also distinguished between types of reader. In the mid-fourteenth century, for example, Dominicus de Clavasio differentiated between the *mensor geometrie* and the *mensor laicus*, explaining in his *Practica Geometriae* that the difference between them lies in 'that what the lay measurer knows how to measure by going and dashing around the sides of a field with his rods and cords, the geometric measurer will know standing still by mental reflection or by drawing lines.'[73] It would appear, then, that there were more learned measurers at work as opposed to some whom Dominicus, like Leonardo before him, considered less sophisticated in their approach, with certain surveyors choosing to calculate distances by walking them and others having the capacity to use more scientific methods. Later still, in the *Geometria Culmensis*, there is yet another distinction made, between *mensores literati* and *mensores layci*, 'literate' and 'lay' measurers, that would seem to equate with Clavasio's *mensor geometrie* and *mensor laicus*.[74]

The practical geometry texts thus reveal a world of different kinds of measurer, some more learned in matters of geometry than others. That these 'gromatic' texts were being written to be read by surveyors is in itself significant. They also reflect the roles and activities of surveyors, which were both urban and rural in nature. This raises the possibility that these differences in expertise and practice were reflected in the outcome of the surveyors' work itself, that is, in the layouts of urban landscapes. The more perfectly geometrical forms of new towns such as Grenade-sur-Garonne and Terranuova may thus have been the work of 'learned' measurers, while those other not orthogonal in form were perhaps the work of measurers of 'lay custom'.

There was therefore a variety of different levels of 'town planning' and a range of 'town planners' in the Middle Ages; from those who led the process, perhaps a king or one of his lords, or the governing body of some city, to those who oversaw the process, the appointed men, not necessarily themselves expert in matters of urban design and surveying but knowledgeable nevertheless about administration and management; to those who actually did the work, drawing up a design and implementing it on the ground by laying streets, marketplaces and building plots, fundamental tasks done either by the same person or by more than one, but all ultimately answerable to 'the founder of the city', they who initiated the work. Whatever the details of this process, geometry was key to the founding of new towns (and in changing existing urban landscapes), in both urban design and surveying.

For newly founded towns, the end of this process was marked in effect by the leading authority issuing a charter setting out the privileges and customs of the place. This 'seventh day' that was the end of the town's creation, the grant of its charter, was an auspicious occasion and symbolically important for all, being marked in the case of the *bastides* by the raising of a pole (*palum*) in the marketplace and by the customs of the town being read out aloud to all those gathered there.[75] With this the town was formally and legally founded.[76] From then on it was down to the town's inhabitants to make of it what success they could.[77]

The cosmogenic parallels between founding a town and founding a world are thus plain, not only in the process of urban formation but also in the use made of geometry to fashion built form. The town was cosmologically symbolic, for embedded in its making was the making of the cosmos, linking Creator with created. The basis of this analogy between urban and cosmological foundation was to be found in scripture as well as the canonical texts of medieval theology and science. All justified the divine authority of the king (or lord), and his place in the celestial hierarchy, while at the same time it was the sovereign that guided the formation of new towns, either directly or indirectly. In this sense urban foundations were microcosms of cosmogony, imitating the work of God and reenacting the formation of the cosmos. They served to instil on earth the universal order of things, linking symbolically those who made new towns with the supreme architect who had made the world. The geometries used in urban design and planning formed a link between the shape of the world and shapes of urban landscapes.

FOUR

MEASURES OF MEANING

PRACTICAL GEOMETRIES AND MEDIEVAL COSMOGRAPHIES

Throughout the Middle Ages the geometrical knowledge drawn upon by those who were creating new urban landscapes was also a source of cosmographical knowledge, the two being entwined in a long tradition of writing about the universe, its order and origin. The significance of this tradition lies in that it gave cosmological meaning to geometrical knowledge – that is, it linked the practical users of geometry, the architects and surveyors of the twelfth and thirteenth centuries who were using it in their creative work, with the divine work of the world's Creator himself. To see this requires us to look further at various *practica geometriae* texts of the high Middle Ages, specifically their content and purpose, as well as their antecedents.

Practical geometry consisted usually of three parts. Hugh of St Victor sets these out in his *Practical Geometry* of the early twelfth century.[1] 'There are three kinds of practical geometry: altimetry, planimetry, and cosmimetry', he explains, in which altimetry 'studies what is high or deep' and planimetry 'occurs for the extension in the plane', while cosmimetry 'gets its name from the cosmos . . . and is measurement of the world.'[2] To learn about geometry for practical purposes meant also learning about the cosmos, for the cosmos is geometrical. Thus Hugh tells the reader that 'the entire world sphere surrounds earth, set as a point in its centre . . . [and that] if we draw straight lines from its middle into the four quarters out to its circumference, we divide the horizon into four parts', and illustrates these principles accordingly (illus. 35).[3] He does all this before

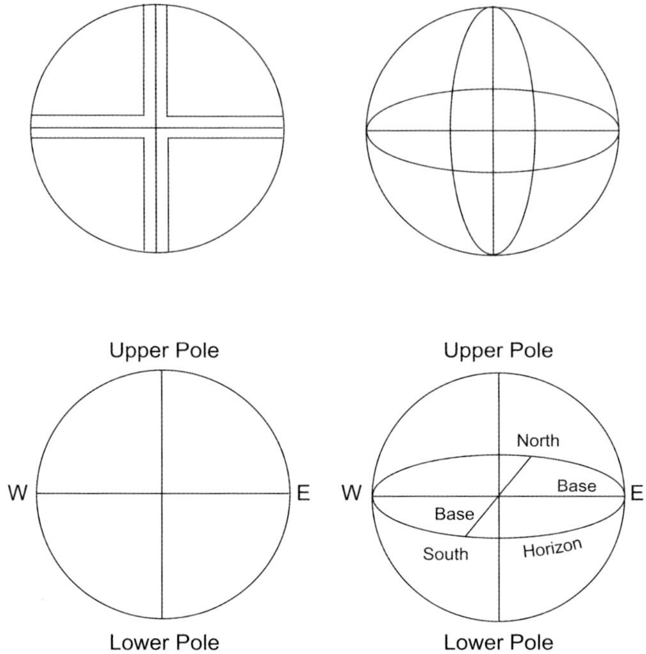

35 Instructional drawings from Hugh of St Victor's *Practica geometriae* (twelfth century): 'We divide the horizon into four parts' (top), and 'From east to pole there will be a triangle . . ' (below).

attempting to discuss the three kinds of geometry, for geometrical understanding proceeded from the measurement of the earth, 'geo-metry'.

The earth was central to geometry, and the earth was central to the cosmos. So while Hugh's *practica geometriae* dealt with the practical problems of surveying, such as different ways of determining heights, recognizing as he does that there were 'less skilful surveyors' among his audience, at the same time it is also an account of the order of the cosmos, 'the physical world', with the earth at the centre of a celestial sphere, drawing upon earlier cosmographies, such as Macrobius' *Commentary on the Dream of Scipio*.[4] Geometry made the world and it was this same geometry that was being used to make urban landscapes.

The connection between 'practical geometry' and cosmography continued to be made in *practica geometria* texts well after Hugh of St Victor was writing, during the twelfth century and through to the fourteenth. In the late twelfth century, for example, the familiar division between altimetry, planimetry and cosmimetry (*crassimetria*) is used by the anonymous author of the French *Artis* text but added to with a fourth section on geometry and astronomy.[5] This is then subsequently repeated by the *Pratike de Geometrie*, a vernacular derivative

36 The '*Psalter map*', thirteenth century.

37 'The Angel shows the New Jerusalem to St John', illumination, ninth century.

38 'The Heavenly Jerusalem', illumination, ninth century.

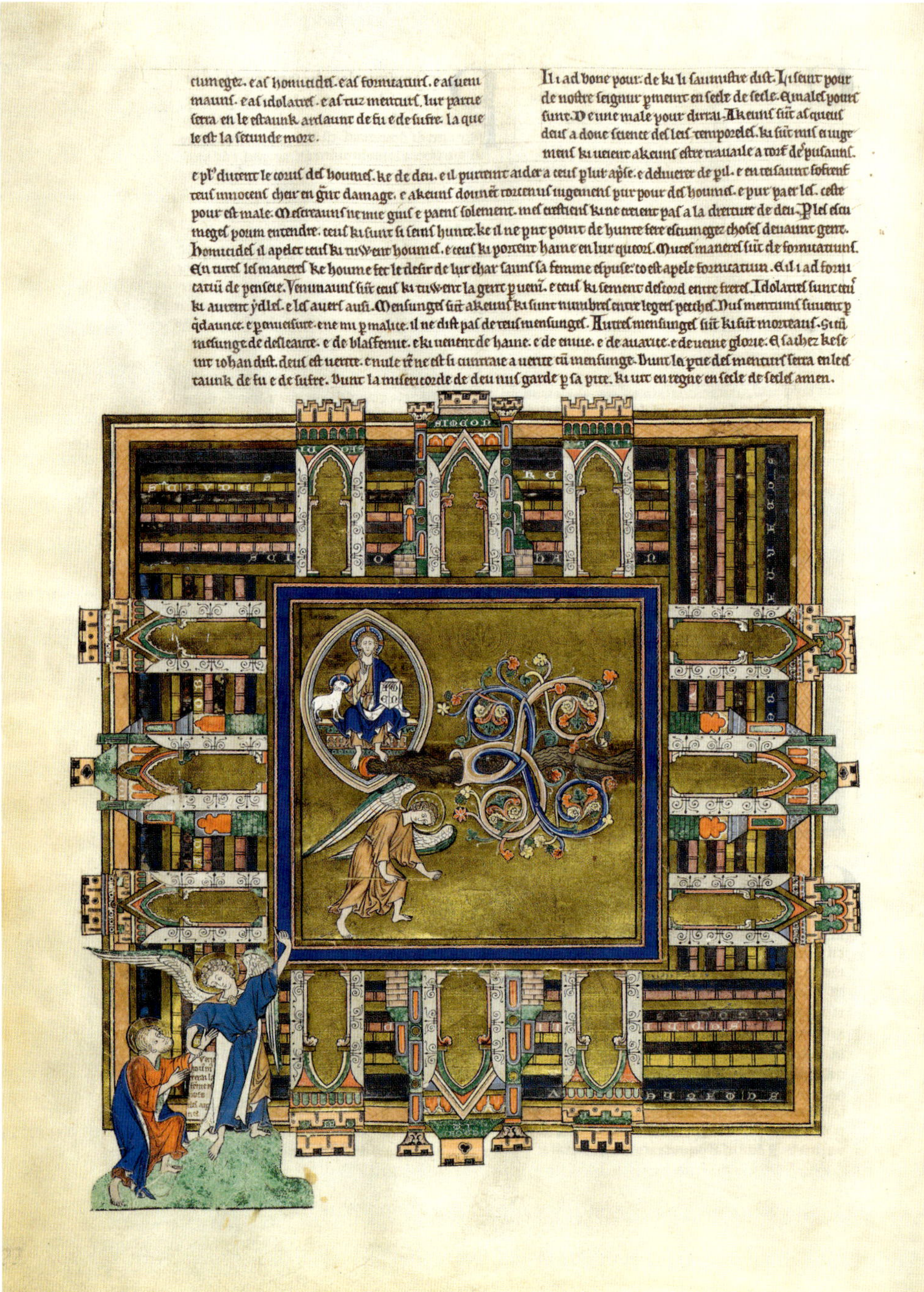

39 The Heavenly Jerusalem, illumination from the Trinity Apocalypse, London, *c.* 1255–60.

40 The Heavenly Jerusalem, *Liber Floridus*, twelfth century.

41 'Crusader map' of Jerusalem.

42 Bristol, *c.* 1480, from 'The Maire of Bristowe is Kalendar'.

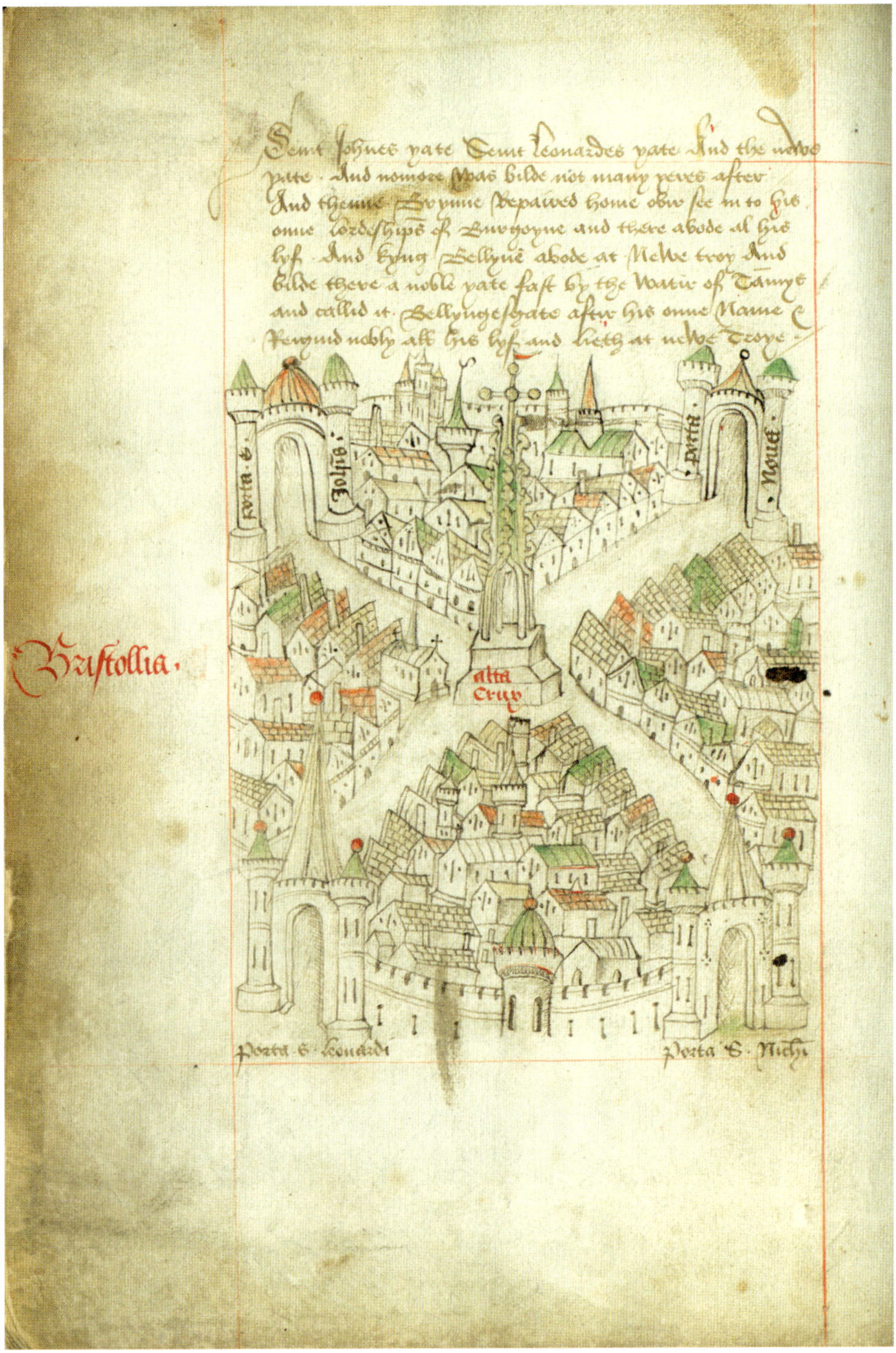
Seint Johnes yate Seint Leonardes yate And the newe
yate. And nomore was bilde not many yeres after
And thenne Brynne departed home ovir see in to his
owne lordeshipe of Burgoyne and there abode al his
lyf. And kyng Bellyne abode at Newe troy And
bilde there a noble yate fast by the Watir of Tamyse
and called it. Bellyngesgate after his owne Name
Regned nobly all his lyf and lieth at newe Troye.

43 Trajectory of the planets, ninth century.

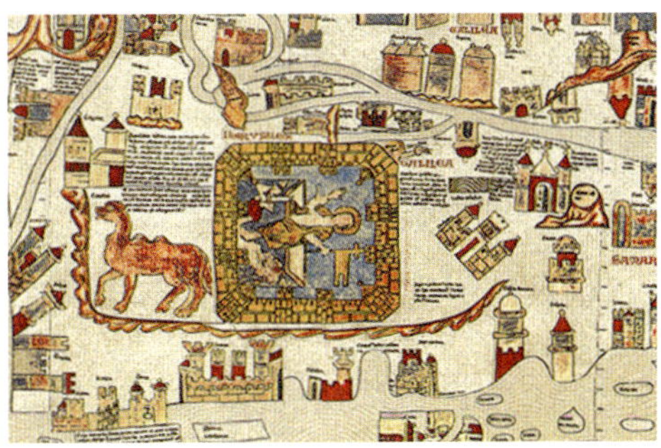

44 The Ebstorf *mappamundi*, *c.* 1234, and its depiction of Jerusalem (original lost).

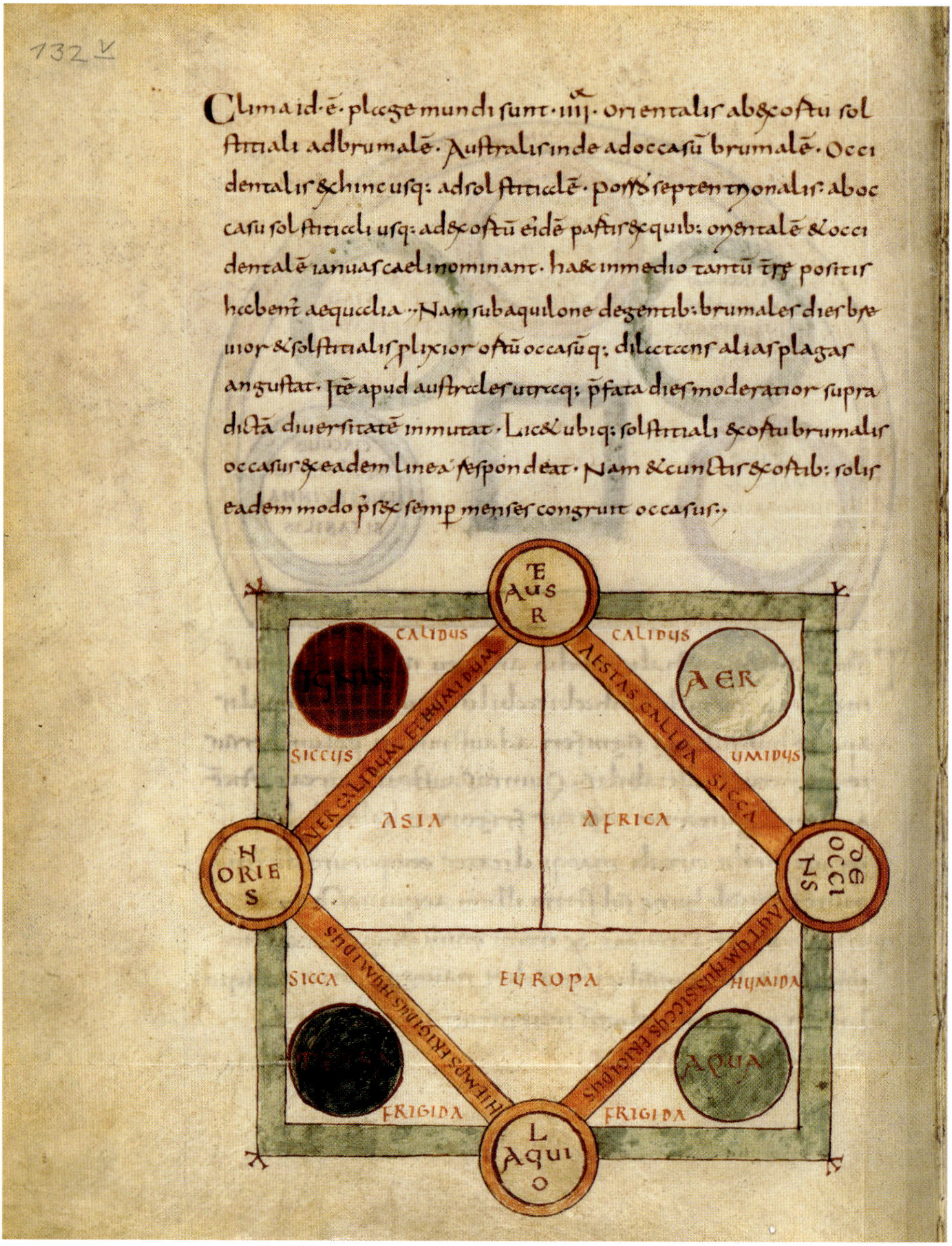

45 Bede's *De Natura Rerum*, ninth-century manuscript.

46 Isidore of Seville, *De Rerum Natura*, late eighth century.

47 Frontispiece of a Bible Moralisée, God as *artifex principalis*, mid-thirteenth century (French).

48 Thirteenth-century miniature.

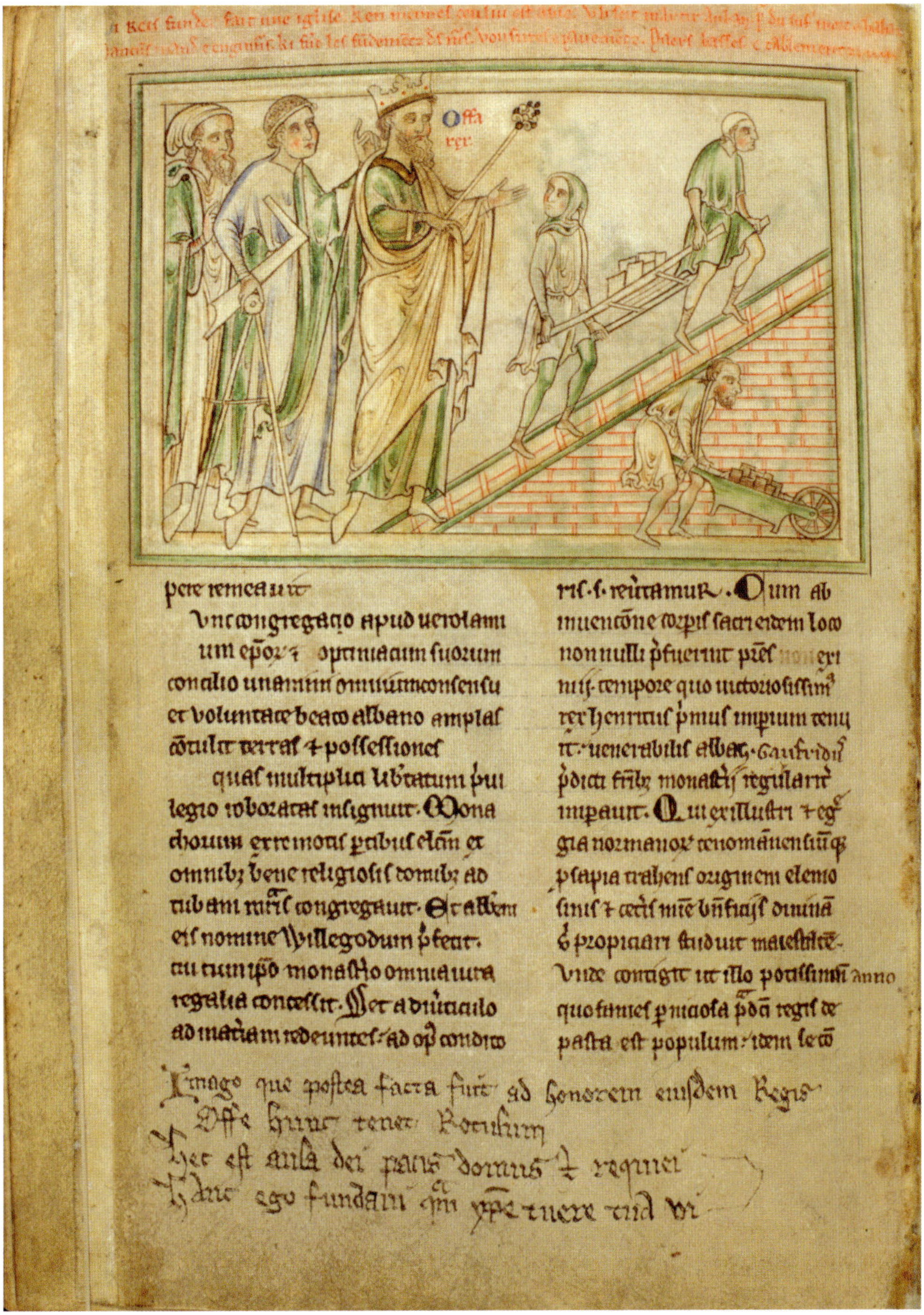
Offa rex

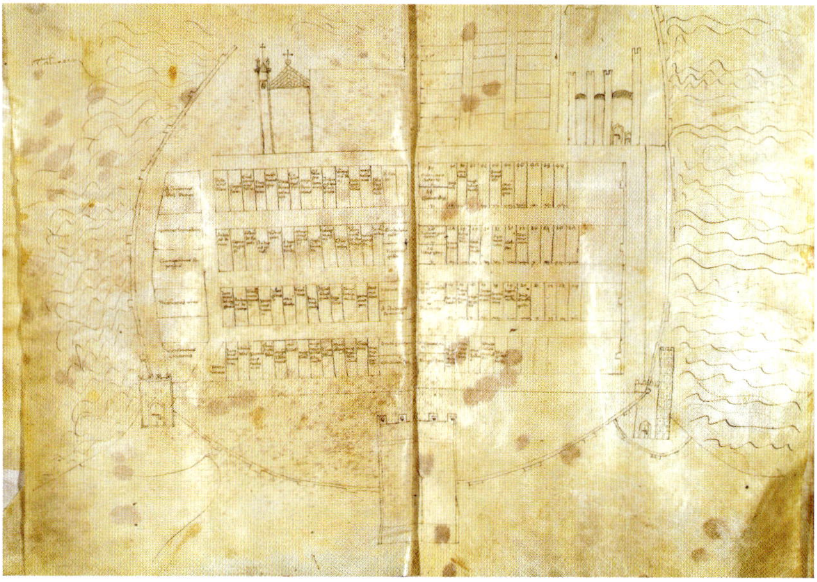

49 Talamone (Italy), plan of the new town, *c.* 1306.

50 Cross-shaped limites extending out from a walled city, *Corpus Agrimensores Romanorum*.

51 Cross of the quartered earth inscribed within a circular heaven in Recension A of the *Corpus Agrimensores Romanorum*.

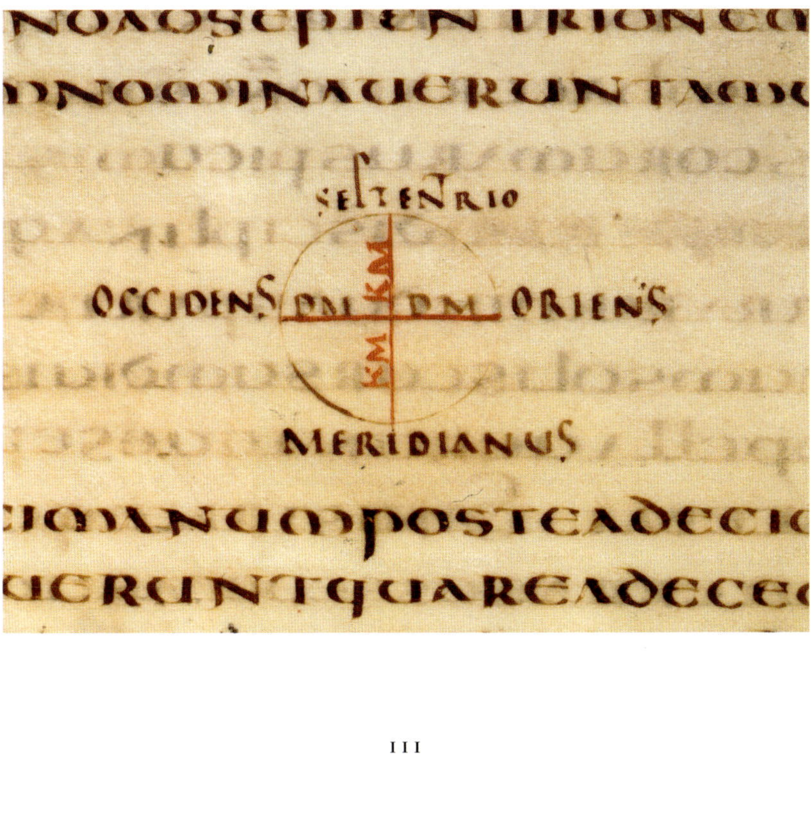

52 Quadrant, fourteenth century (English).

53 'The Chaucer astrolabe', *c.* 1326 (English).

of thirteenth-century date.[6] The *Artis* and *Pratike* texts both draw geometry and cosmography together, both in their use of examples (for example, 'to find the ratio of the volumes of the sun, the moon and earth'), and also in their use of methods of measurement, using an astrolabe, for instance.[7] A practical geometry text thus not only had material for potential use in urban planning, for example on how 'to infer the number of houses in a round city', but also material that informed the reader about the geometry of the wider world.[8]

Geometry's cosmological basis was plain to see therefore, and so too were the cosmogenic dimensions of geometrical figures being used in the plans of certain medieval towns and cities. Likewise in the thirteenth century, this connection is also to be seen in more 'academic', cosmographical texts that dealt with geometrical knowledge, such as the *Sphere* of Sacrobosco, and a *Commentary* made on it by Robertus Anglicus, as well as in the *Treatise on the Quadrant*, a 'treatise quite similar to the practical geometries'.[9] In the *Sphere*, for example, considered to be 'the clearest, most elementary, and most used textbook in astronomy and cosmography from the thirteenth to the seventeenth century', Sacrobosco begins his work by describing the sphere as a geometric figure, in Euclidian terms, before going on to examine the world of spheres that made up the Christian cosmos.[10]

These texts, and the ideas contained within them, were circulating during the twelfth and thirteenth centuries when much of Europe's urban expansion was taking place. It is very likely that those involved with shaping these urban landscapes picked them up. This would include architects and measurers for whom the *practica geometriae* were seemingly written,[11] as well as administrators, such as clerks working in royal households or for municipal governments and who, as Shelby suggests, were taught the *quadrivium* (which geometry and astronomy were both part of) in their formal education at school or university.[12] They would be known too by the patrons of founded towns themselves.[13] In all, those individuals who we know were creating new towns in the Middle Ages could not easily have overlooked the cosmological dimensions of geometrical knowledge which informed their thinking and their work.

Theoretical and practical geometries

In the Middle Ages, 'theoretical geometry' basically meant Euclidian and Pythagorean geometry, an abstract mathematical geometry.[14]

In which case, why was it seen to be linked to Christian conceptions of the cosmos? The answer to this lies in geometry's place in scholastic learning as part of the *quadrivium*.[15] By the early twelfth century, thanks to Adelard of Bath, knowledge of Euclid in Europe was greatly advanced through a translation into Latin of Arabic copies of the *Elements*.[16] Before this, what was known of the *Elements* was partial, being based on a 'translation of Euclid prepared by Boethius', which formed one part of his four treatises on geometry, arithmetic, astronomy and music that made up the *quadrivium*.[17] Boethius embedded Euclidean geometry into a tradition of Christian cosmological thought that was fundamentally Neoplatonic in outlook and which was at the heart of medieval learning and scholarship. Boethius felt that 'it was impossible to achieve the summit of perfection in the disciplines of philosophy unless one approached this noble wisdom by a kind of fourfold way', a belief which he put to work in the dialogue of his *Consolation of Philosophy*, his version of Neoplatonic cosmography.[18]

Through Boethius, then, a theoretical, Euclidian geometry was not only known about in the earlier Middle Ages but was united with (Neoplatonic) cosmography. Boethian geometry with its Euclidian material was also merged with other cosmographic and geometric knowledge through later texts that have been attributed to Boethius, such as one of eleventh-century date, but which are now known to be spurious.[19] This, as well as other 'unauthentic' (pseudo-) Boethian texts of ninth- through to fourteenth-century date, contained 'much practical matter on surveying' as well as 'fragments of a Latin version of Euclid thought to have been the one prepared by Boethius'.[20]

These 'gromatic texts' are known by the title *ars geometriae et arithmeticae* and as Ullman sees them they are 'a condensed Boethius with matter added from other sources', that is, sources which included selections from Isidore, Martianus Capella and Cassiodorus, as well as astronomical treatises and excerpts from Gerbert's *geometriae*.[21] For the readers of these 'excerpt manuscripts',[22] geometry was again combined with cosmographic knowledge provided by Neoplatonic sources, so while the *Elements* was itself not by any means a cosmographic text, its place in a (pseudo-)Boethian *corpus* of 'gromatic texts', widely known and read as a source of both geometrical and cosmographical knowledge throughout the Latin West, provided such an association. The conceptual link between Euclidian geometry and cosmography was not broken once Boethian versions of the *Elements* were superseded by Adelard's translation. It continued to be forged in the *practica geometriae* of the later Middle Ages.[23] Being a source used in the *quadrivium*, a 'theoretical', Euclidian geometry thus rested alongside what was known of the Neoplatonic world, its order and its making, and the abstract geometries that they both shared were an important part of medieval learning well before the twelfth century and before Adelard's new translation of the *Elements*.

By the twelfth century, two types of geometrical knowledge fed into gromatic texts, and thus into the minds and practices of their assorted readers. Both related geometry to the cosmos. One of them was 'theoretical' geometry, the other 'practical geometry',[24] a distinction drawn by Hugh of St Victor in the *prenotanda* of his *practica geometriae*, where he notes 'geometry is either theoretical (speculative) or practical (active)', defining the first by saying it 'uses sheer intellectual reflection to study spaces and intervals of rational dimensions', and the second as geometry which 'uses instruments'.[25] Similarly, in his *De divisione philosophiae* of around the same time, Dominicus Gundissalinus differentiated between 'the artificer of theory', that 'is the geometer, who clearly knows all parts of geometry and can teach it', and 'the artificer of practice', those 'who use [geometry] in working', of which 'there are two kinds . . . namely, surveyors and craftsmen.'[26]

This distinction between theory and practice is echoed later too, in the practical geometries of the thirteenth and fourteenth centuries, by the way they sought to define two audiences of their work, as with those written by Leonardo Fibonacci and Dominicus de Clavasio (see above, chapter Three). By then *both* types of geometry were being presented as two means of approaching problems in the *practice* of measure and survey work, one involving thinking and calculating, the other pacing with rods and cords, whereas earlier, in the twelfth century, the separation of the two kinds of geometry appears to relate more to differences in their intellectual origins and separate kinds of user – the theoretician and the practitioner. Thus Adelard of Bath, writing in the early 1100s as was Hugh of St Victor, could explain the difference by differentiating between the *demonstrator*, whose 'duty . . . is to explain theorems', and the practitioner (*exercitator*), whose 'duty . . . is to measure' using instruments.[27] Both kinds of geometer coexisted in the twelfth century, and drew their knowledge from differing traditions, one from the Greeks, the other from the Roman surveyors, the *agrimensores*. Fragments of each were being brought together in gromatic works such as Hugh of St Victor's *Practical Geometry*,[28] and earlier texts on geometry of the tenth and eleventh centuries, notably the *Geometria* of Gerbert d'Aurillac (Pope Silvester II),[29] which both had practical aspects to them but were also both studied by schoolmen in the *quadrivium*.[30]

What begins to change during the twelfth century is the target audience of the written practical geometries, some of which seem to be aimed more towards educating Adelard's 'practitioners' and Gundissalinus' 'artificers of practice', giving them the knowledge, for example, to make their measurement calculations more theoretically based or providing them with a new approach for surveying the land. The two traditions feeding into this knowledge and practice of geometry of the twelfth century onwards were *both* related to cosmographic thought, a connection that was perpetuated in the *practica geometriae* texts of the later Middle Ages. The first of them is the pseudo-Boethian 'theoretical' geometry, mentioned above, while the second came from those gromatic texts based upon surviving manuscripts of Roman surveyors, the written works of the *agrimensores*.

The legacies of agrimensores

A set of Roman surveying manuals, the *Corpus Agrimensorum Romanorum*, dealt with matters of land surveying and urban planning, as well as aspects of cosmography.[31] The importance of these texts for us here lies in how their material on knowledge about the cosmos, such as the harmony of the spheres, was embedded in geometrical knowledge of a practical nature for use by medieval land surveyors, for example how to measure unknown distances.[32] The nucleus of the *Corpus* goes back to the first or second centuries AD, and 'served in the training and subsequent practice of professional surveyors and other students of higher learning.'[33] The manuals were subsequently copied and collated between the fourth and seventh centuries in Italy, and gradually had further cosmographic commentaries and gromatic texts added.[34] The 'genealogy' of the manuscripts is summarized in illustration 54. The *Corpus* continued to be copied into the ninth and tenth centuries, for example at centres of learning such as Bobbio in Italy, as well as Corbie in France and Aachen.[35]

These important texts have received attention from historians for over a century, but on the whole their studies have examined what the manuals reveal of Roman surveying practices.[36] The survival of the texts into the Middle Ages and their place in the *quadrivium* has also received attention,[37] but less consideration has been given to the influence of the *Corpus* manuals on medieval surveying and architec-

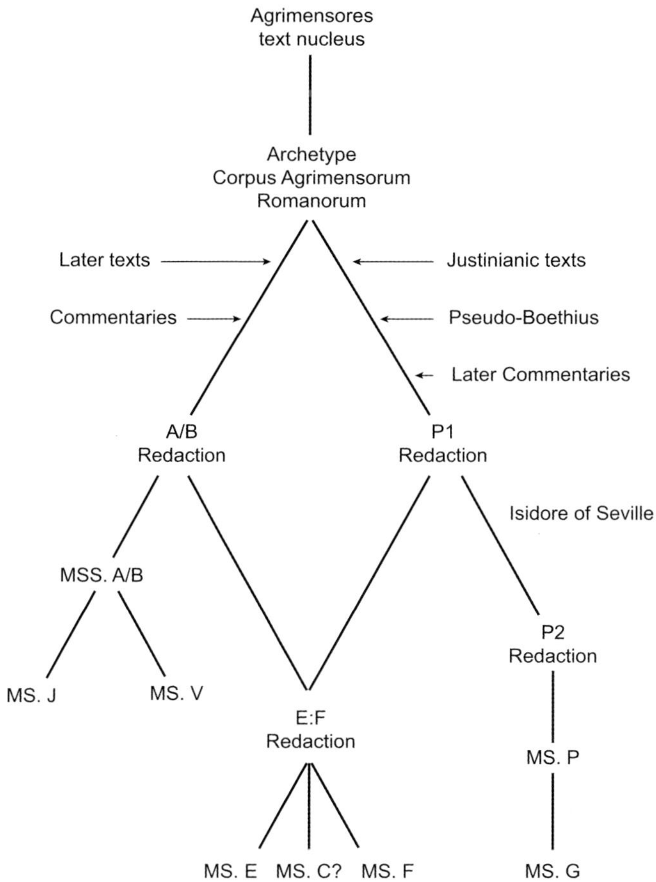

54 A genealogy of the *Corpus Agrimensores Romanorum* texts.

ture. Homann notes that the gromatic works of Gerbert and Hugh of St Victor each owe something to the *Corpus*,[38] while Ullman has considered its place in the 'excerpt manuscripts' and the *ars geometria*.[39] However, Ullman argues that the *Corpus* texts were not studied for their practical surveying content, at least in the Carolingian period in northern Europe, and suggests that instead they were used as sources of knowledge for the teaching and study of geometry as an academic subject in schools.[40] The texts could have had two audiences, of course, and gained wider, practical usage in part because their contents on surveying problems would have been helpful to those surveyors and craftsmen whom Dominicus Gundissalinus and Hugh of St Victor said were engaged with 'practical geometry', and in part because the demonstrations used in the manuals are also to be found in the later *practica geometriae* texts.

The survival of these surveying manuals through the Middle Ages, and their circulation as copies made in the Latin West, thus could have provided not only those who wished to know how to resolve surveying problems with information but may also even have owed their survival to practices of surveyors continuing the Roman tradition.[41] By reading of Roman surveying practices, a connection was once again being made between geometry and cosmographic knowledge (as it was with the pseudo-Boethian texts which were likewise part of the *ars geometriae*).

Through the gromatic texts of the *Corpus* manuals, therefore, understanding the wider world was linked to the world of those who were using practical geometry. Whether they were dealing more with theoretical geometry or practical geometry, these gromatic texts combined cosmographic and geometric knowledge. These, then, were the inherited cosmographic geometries that the antique world handed on to those who came after, the schoolmen and the surveyors of medieval Europe. This tradition can be traced through the Middle Ages through copying and circulation of texts, through to the twelfth, thirteenth and fourteenth centuries, by which time the various *practica geometriae* were specifically addressing problems that surveyors and architects confronted in their work (see chapter Three).

The content of these gromatic texts points not only to a marriage of cosmographic and geometric knowledge, in which geometry is seen to be cosmologically symbolic, but also to a practical use of this geometry in the work of artificers who were involved with measuring and measurement of the land. The *practica geometriae* which forged both theoretical and practical geometry were not written solely for scholarly study, they were meant for those who required practical instruction in measurement, in altimetry, planimetry and cosmimetry. Anyone reading these gromatic texts and using them in their work could not have failed to ignore the cosmological dimensions in the knowledge and use of geometry. It was, after all, the same geometry that gave form to the world and which had formed it, as contemporary depictions of God the geometer showed.[42]

To see how surveying was symbolic of both cosmology and cosmogony means moving from geometry as a body of knowledge, and beyond the conceptual links between cosmographic and gromatic texts *read* by surveyors and architects, to consider more the cosmological symbolism of surveyors' practices and the instruments they *used* in their work. That measurement and measuring had such symbolism in the Middle Ages is evident from descriptions made of surveying in the gromatic texts as well as in scripture.

SURVEYING INSTRUMENTS AND COSMOLOGICAL SYMBOLISM

The *Corpus Agrimensores* texts and their gromatic derivatives provide one reason why surveying had cosmological symbolism in the Middle Ages. For the Romans the very act of surveying was itself cosmically orientated. Campbell has studied the *Corpus* texts for what they reveal of surveying practices in ancient Rome, particularly the methods surveyors used to choose a site and found a colony, creating an urban settlement.[43] This process began with setting out two intersecting lines (*limites*), the *kardo maximus* and *decamanus maximus*, each orientated cardinally through the use of geometry, which then formed the basis of the colony's urban layout of streets, and outlying property and field patterns and roads.[44] The *limites* 'were the essential framework upon which all land measurement was based', and to lay them out surveyors used 'the *groma* or *ferramentum*, a kind of cross-staff used for plotting straight lines and measuring right angles.'[45] These surveying practices were also illustrated in the *Corpus* texts, cities being shown in schematic but geometrical form within a geometrically ordered landscape (illus. 50).[46]

Through the continued copying and circulation of the Roman *Corpus* texts some in the Middle Ages would get to know about the practice of making *limites* and their purpose in founding new urban settlements.[47] The cosmological dimension of this procedure would also be known, for the texts make it clear that what the surveyors were doing had a ritual, cosmogenic basis. Rykwert has discussed 'the cosmic implications of surveying in the *agrimensores* texts', and argues they 'are a "rationalized" and weakened survival of the [earlier] Roman-Etruscan belief in the sacredness of the land titles and boundaries.'[48]

In the *Corpus*, Hyginus makes it evident that orientation of the *limites* had cosmological significance: 'among all the observances and practices of measurement the most distinguished handed down to us is the establishment of *limites*. For it has its origin in the heavens and its legacy is timeless . . . For *limites* are established on a principle not unassociated with cosmology, since the *decumani* are aligned according to the course of the sun, the *kardines* from the axis of the earth.'[49] Laying out the *limites* was thus cosmically inspired, quartering the ground as the earth is itself quartered in an act of foundation.[50] Frontinus put it more succinctly: '*kardo* gets its name because it is drawn from the pivot of the universe'.[51] Frontinus gives thorough instruction on how to use the surveyor's *ferramentum*, the instrument used to set out on the ground the straight lines and right angles extending from the axes of the *limites*, as well as indicating to readers what the cosmological dimensions of surveying were, and which cardinal points should be used in setting out the *limites*.[52] The use of the gnomon to orientate according to the heavens is also explained, for example in the Hyginus text.[53]

The *Corpus* illustrated these cosmological principles of surveying, too, particularly the cross-shaped *limites* extending out from a walled city, and the cross of the quartered earth inscribed within a circular heaven (illus. 50 and 51).[54] So medieval readers of the *Corpus* texts were provided with evidence of the cosmological basis of surveying, as well as an account – both written and visual – of how these practices were used in Roman urban planning. Through orientation and squaring the ground, heaven and earth were thus drawn together.

Linking heaven and earth

The *Corpus* texts thus made clear the cosmological foundations of terrestrial surveying. Through their circulation during the Middle Ages, a time when many of Europe's new towns were being created, surveyors who knew these texts would have been aware of this connection. How far medieval surveyors were actually being influenced by Roman *agrimensores* texts, surveying techniques and instruments is a question that has actually received scant attention among medievalists, however. On the basis of the very survival of the copied *Corpus* manuscripts into the Middle Ages some influence is at least likely, but were medieval urban surveyors using a *ferramentum* or *groma* to fix their street intersections at right-angles, and was a gnomon used to calculate the orientation of a town's layout, as the *Corpus* texts instructed? If so, then the cosmological meanings attached to these instruments, and the cosmogenic practices associated with using these instruments, probably survived through into the Middle Ages.

The Roman agrimensorial tradition has been suspected as an influence on medieval town planning, for example through the presence of quartered layouts and square-modules of planning that look similar to the *limites* of the Roman *colonia* (see illus. 18, 19 and 50).[55] Cartographic historians have also noted the survival of Roman agrimensorial methods in the scaled plan of the Carolingian monastery of St Gall,[56] drawn in the early ninth century, a further indication that Roman surveying practices were continuing into the Middle Ages in the field of architecture, which raises the likelihood that there were those who were capable of measuring and setting out new towns in the same manner, too, towns such as Alfred's *burhs* of Wessex perhaps (illus. 11). In what ways Roman surveying knowledge and practice might have been used and made manifest on the ground still requires pinning down, however. There is of course textual evidence for such continuity within the gromatic books of the later Middle Ages, the *practica geometriae*.

Surveying instruction in the *practica geometriae* derives in part from the tradition of the Roman *agrimensores*. Dilke notes that one of the *Corpus* texts, the earliest manuscript, the Arcerianus of the sixth or seventh century, was at the monastery of Bobbio at the end of the Middle Ages and that 'it may well have been there in the tenth century, when Abbot Gerbert, later Pope Silvester II, studied the works of the *agrimensores*' for writing his treatise on geometry.[57] Gerbert's *geometriae* was highly influential later on during the Middle Ages, as witnessed, for example, by Hugh of St Victor's *Practical Geometry* and the *Artis* text, both from northern France, both of the twelfth century.[58] Gerbert, for instance, provided instruction on how to use a gnomon 'in connection with the setting out of the *cardo* and *decimanus* [*sic*].'[59] Which *Corpus* texts he was actually drawing upon is not certain. Bubnov suggested it was not the Arcerianus, though others have noted 'similarities between the Arcerianus texts and the mathematical writings of Gerbert', in which 'quotes are found from the Hyginus text and other gromatic studies.'[60] The gnomon was widely discussed by Roman sources, not just the *Corpus* texts and not just the Arcerianus manuscript.[61] It is shown to be used by the Hyginus text in particular for orientating the principal *limites*, both in the written description and also in the accompanying illustrations, one of which, although technically incorrect, shows this purpose quite explicitly (illus. 55).[62] The gnomon is simply a stick, staff or pole.[63] Its cosmological symbolism lay in its connection between the motion of the sun and the laying out of the *limites* on the ground, in the founding and planning of a *colonia*.[64] This ordered, quartered form of the principal *limites* – the *cardo* and *decumanus* – thus rested

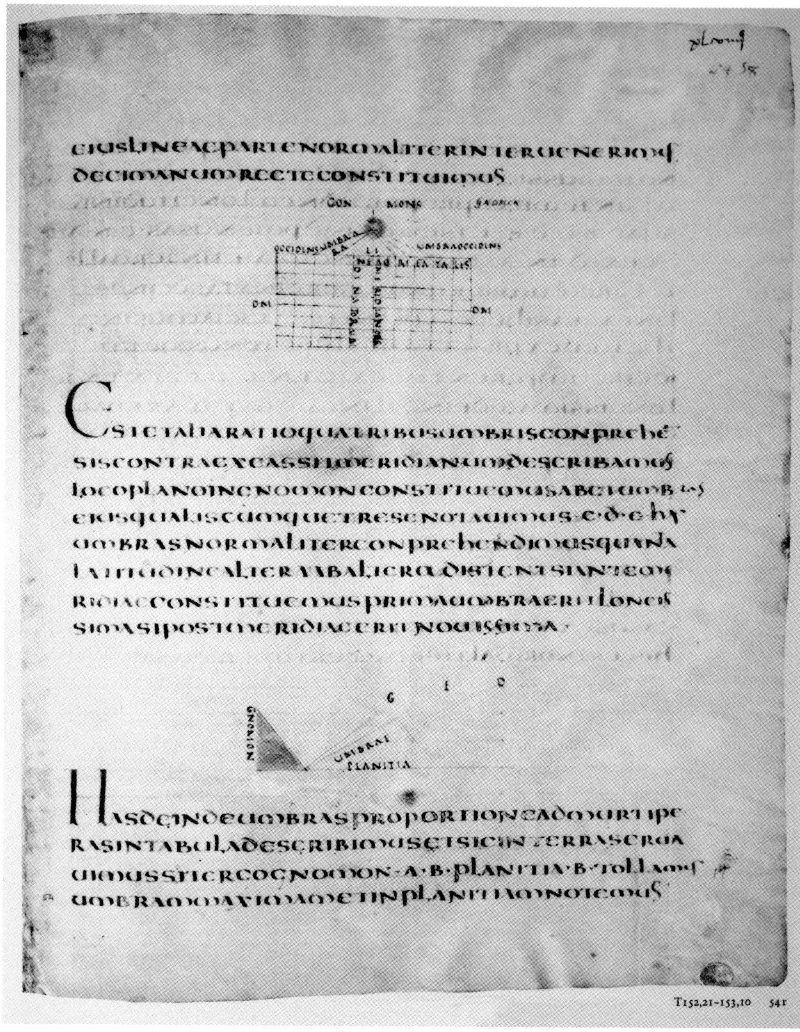

55 Using a gnomon to orientate the principal *limites* according to Recension A of the *Corpus Agrimensores Romanorum*.

on the correct use of the gnomon to orientate correctly to the heavens. This is what Gerbert was advising on.

Considering that so many layouts of medieval towns are cardinally orientated and quartered with principal cross-streets, it is perhaps no wonder that the suggestion has been often made that the practices of the Roman *agrimensores* were continued by medieval surveyors.[65] A corroboration of this lay in the texts of the *practica geometriae* of the later Middle Ages, where again the use of the gnomon is mentioned. In his practical geometry, Hugh of St Victor, for example, considers the gnomon, though describing its use in calculating solar altitude – a parallel function of the instrument – rather than as a means of orientating, that is, in its practical value in cosmimetry rather than planimetry.[66] The anonymous author of the *Artis* text, who we know drew much from Gerbertian geometry, also makes use of the gnomon, this time in altimetry, but again as a means of determining the altitude of the sun rather than surveying lines on the ground.[67] However, in his *practica geometriae*, Dominicus de Clavasio set out 'how to measure the distance between the eye and a point in a valley with the *instrumentum gnomonicum*.'[68] These texts thus show a continuing usage of the gnomon into the Middle Ages, and knowledge of its cosmological basis and function. The possibility surely exists, then, as Gerbert had himself explained, that the gnomon was used for orientating ground plans to the cardinal points (especially east–west) as well as observing heavenly movements. Indeed this possibility is given added weight by the fact that so many gromatic texts based on the *Corpus* were circulating and being copied after the ninth century, in northern France for example, where Hugh and other authors of *practica geometriae* were working.[69]

While the gnomon is clearly referred to by medieval sources on surveying, the presence of the *agrimensores*' *groma* is more enigmatic. Kiely is adamant that the *groma* was not known by medieval surveyors, saying 'not even Gerbert, who was nearest to the Romans in point of time, makes any mention of the *groma*', and this despite Gerbert's use of the 'gromatic' *Corpus* texts.[70] The *groma*'s practical function, together with its cosmological symbolism in the laying out of the *limites*, was thus conceivably lost on the medieval audience of the agrimensorial texts. Yet, as we have noted, right angles were used in orthogonal layouts of streets in medieval towns, the setting out of which could have been accomplished with accuracy using the *groma*, or a similar four-squared instrument, coupled with the expert advice offered, for instance by Hyginus, with which readers of the *Corpus* would have been familiar. There is an indication that this was in fact the case. A description of an instrument mentioned in the anonymously authored *geometria culmensis* (*c.* 1400) from Pomerania suggests a groma-like device.[71] Simply called 'the cross' (*crucze*), it was 'used for laying out a perpendicular line' in surveying.[72] The presence of such an instrument in east-central Europe, an area where many new towns were being laid out with square-based plans, led Zagrodski to conclude that there was indeed continuity in surveying practice and usage of the groma for town-planning purposes.[73] If a *groma*, or its equivalent, was known and used by medieval surveyors, the cosmological and Christian symbolism of its cross-shape would surely have been apparent, even if they had not themselves read of its cosmological significance from the texts of the *agrimensores*.

Geometric practice and microcosmic symbolism

If the *groma* was not the instrument used by medieval surveyors to set out geometrical town-plans on the ground, one alternative that certainly had suitability was the quadrant (illus. 52).[74] This, together with a closely related innovation, the astrolabe, came not from Roman surveying traditions but from Arabic contacts via the Mediterranean, especially Spain and Byzantium.[75] From the time of Gerbert, the astrolabe was known and used in the Latin West, while the quadrant, based upon the same principles as the astrolabe, came rather later, in the thirteenth century.[76]

Both the quadrant and astrolabe were cosmimetrical instruments, and thus pointed to the heavens, but they were at the same time also used in practical matters of surveying, in planimetry and altimetry, as is made clear in *practica geometriae* of the twelfth century onwards, such as the *Artis*, and of course Hugh of St Victor's, as well as those by others such as Leonardo Fibonacci.[77] The quadrant was discussed in detail particularly by Robertus Anglicus (alias John of Montpellier) in his *Quadrans vetus* of the late thirteenth century,[78] parts of which subsequently appeared in the earliest English vernacular work on surveying, the *tretis of geometri* of the fourteenth century, in an exposition given there on the quadrant's practical use.[79] Other tracts on the use of the quadrant also appeared at around this time: one in a manuscript attributed to John of Sacrobosco, for instance, is sandwiched in 'between his *Computus* and his *Sphere*', the latter a cosmographical work on which a *Commentary*, written in about 1271 by Robertus, also made the cosmimetrical role of the quadrant clear.[80] The quadrant's astronomical functions thus gave it a cosmological significance, linking surveying practice with the macrocosm. The same is true for the astrolabe.

The astrolabe had received attention by Gerbert, probably because of the time he spent in Spain in the 960s.[81] It was discussed at length by Hugh of St Victor in his *Practical Geometry* that owes so much to Gerbertian texts, in which he notes 'the astrolabe is surely the surveyor's most important instrument' (illus. 53)[82]. The various astronomical and practical uses of the astrolabe are illustrated and discussed in the *Artis* text, also inspired, it seems, by Gerbert, and there it is noted, significantly, that the instrument may be used for measuring over the ground, both in height and in distance.[83] This approach meant that a surveyor could avoid having to measure using fixed rods or flexible cords, and so improve their speed and accuracy of distance measurement and setting out (see chapter Three).[84] Some variation in knowledge and usage of the astrolabe in practical surveying work in the thirteenth century is suggested by its *absence* from the text of Villard de Honnecourt's *Sketchbook*,[85] an absence that reinforces the idea that there were different kinds of surveyors working on laying out the new towns of the Middle Ages, not all of whom shared the same geometrical knowledge and practical skills.

Along with their practical utility, these instruments had symbolism. In terms of their geometrical shape, the astrolabe (illus. 53) is round like the cosmos, while the quadrant has the shape of the quartered world (illus. 52).[86] Handling them was to hold the world *in parvo*, while using them to create order on the ground mirrored the observed order of the created cosmos. The dual role of these surveying instruments as a means of observing and measuring the heavens, coupled with their value in the measuring work of the surveyor on the ground, might indicate that (as with the 'cross') their users were not oblivious to the microcosmic resonances of the form and function of these devices. Symbolically, then, the astrolabe and quadrant inscribed heavenly meaning into the material forms they created, built forms perhaps.

Even if architects and surveyors were not directly influenced by the *Corpus* and its later derivatives, the cosmogenic and cosmological connotations and symbolism of their instruments would surely have been known to them through what they learned from reading Scripture. The Bible itself refers to the work of architects and measurers, and mentions instruments used by them, all of which provided scope for both literal and mystical interpretations. By the twelfth and thirteenth centuries, geometry was being used in exegesis, for example by Hugh of St Victor, Richard of St Victor and Roger Bacon, whose concern was with using their geometrical knowledge to explain biblical passages such as Ezekiel's vision: 'it is not possible for the literal sense [of Scripture] to be known unless a man have these works depicted, but more so when they are pictures in their physical forms.'[87] The 'geometric description of the ark' given by Hugh includes 'much more geometry than he needs strictly for exegesis', Victor points out.[88] The same could be said for Hugh's disposition on the use of allegory in his *Didascalicon*, where he explains in detail, using analogies drawn between the work of a mason and their technique of laying foundations and courses of walls, how 'divine scripture is like a building'.[89] Thus, noting how the mason 'stretches out his string in a straight line' and 'drops his perpendicular', Hugh alikens raising a building to the process of Biblical study 'to construct a spiritual building'.[90]

The use of geometry in literal and mystical interpretations of Scripture was evident also in manuscript illustrations, most notably in those of the eleventh to the fourteenth centuries that commonly depicted God with dividers or compass in hand creating the universe (illus. 47).[91] Through them the compass as a geometrical instrument is given particular microcosmic symbolism, promoting the (Neoplatonic) idea of an 'architect God', and hence also cosmogenic symbolism in the work of the medieval architect (illus. 9 and 48).

The compass traces a circle, and was used, as noted earlier, to draw the geometrical forms of new towns such as Talamone. It is thus an instrument of architectural and urban design (illus. 49).[92] In depictions of God creating the cosmos he uses the compass in order to trace its circular form, and carve order from chaos, but as Friedman points out, the most likely sources for this imagery do not come from Christian commentators but rather derive from Jewish scholars working in the Latin West, such as 'Rashi, writing in Troyes in the eleventh century', who 'is well known for studding his scriptural commentaries drawn from daily life'.[93] Thus in the Old Testament, in Proverbs, the prophet Solomon tells us that 'when He prepared the heavens', God 'set a circle (*gyrus*) on the face of the deep'.[94] In his commentary on Proverbs, Rashi notes that 'the meaning is circumference, or what is marked out, as in Isaiah 44:13, "and he marked it out with a compass"'.[95] So although the 'architect God' may well have been inspired by medieval Neoplatonic sources, such as William of Conches' *Commentary on the Timaeus*, and Daniel of Morley's *Liber de naturis*,[96] the compass as an instrument used by God in Creation to trace out cosmological form would seem to owe more to medieval Hebrew sources. Both were being picked up by manuscript illustrators. From the New Testament there is of course another source for seeing God as a creator and as an architect. Hebrews states Abraham 'looked for a city with foundations whose builder and maker is God', a phrase which if interpreted mystically might suggest a city, a 'world', created by an architect-God, for whom the compass would have been the most obvious, pertinent geometrical instrument and cosmological symbol.[97] The cosmological symbolism of the compass was thus widely understood in the Latin West in the Middle Ages, and surely gave earthly users a sense of its symbolic significance and meaning in their own design work.[98] Like the measurer's tools – the quadrant and astrolabe – the geometer's compass imitated a macrocosmic geometry, reproducing on earth what is in heaven.[99]

While Scripture and later commentaries indicate a cosmogenic role for the compass, the principal surveying instrument referred to in the Bible is the measuring reed. The use of a reed to measure is referred to especially, and indeed significantly, in the two visionary accounts of Ezekiel and John, in their revelations of cities of the future, where it acquires divine significance. The prophet Ezekiel describes 'visions of God' in which he reveals an image of a new world and sets out in detail the specific measurements as instructed by 'a man, whose appearance was like the appearance of brass, with a line of flax in his hand and a measuring reed', measurements of the Temple's buildings and walls, five hundred reeds on the north, east, south and west sides 'to make a separation between the sanctuary and the profane place'.[100]

The measurements and the measuring using a reed are further accompanied by instruction on the order of the buildings and surroundings, particularly their orientation to cardinal points, and the site and size of 'a profane place for the city, for dwelling, and for suburbs', where 'the city shall be in the midst'.[101] The city is four-square, being 4,500 reeds on each of the sides, with suburbs beyond.[102] The regular form of Ezekiel's city and its metrological basis using linear measures were thus associated. That geometry was used in the Middle Ages to try to make sense of the 'physical forms' of Ezekiel's vision is made clear by Richard of St Victor and, later, Roger Bacon.[103] For the surveyors at this time the equivalent to the biblical reed was the rod, of fixed length, while the 'line of flax' had become a cord or chain, a more flexible measuring line than a rod made of wood (illus. 34).[104] Despite the changes in nomenclature, Ezekiel and medieval surveyors principally used the same instruments for ordering and measuring a city. Ezekiel's vision furnished these basic and widely used implements with a symbolic significance: they were the means by which to create and lay out a new world.[105]

Through Ezekiel, then, the medieval surveyor not only had knowledge that theirs was a tradition endorsed by scripture, but that their instruments had a divine standing which gave them a status equivalent to that held by architects through their compass. This symbolism of the measuring reed was of course further reinforced in John's revelation of the Heavenly Jerusalem, where the golden reed, provided by God, ordered the celestial city, the 'reed like unto a rod'.[106] Here there is an added significance. The cosmogenic dimension of the new Jerusalem meant that the measuring reed was not simply a divine instrument used to give order to the world to come but a symbol of the cosmogony of the world itself. The measuring reed in the mind of a medieval surveyor could hardly be a more significant Christian symbol of the world to come, and in contemporary Apocalypse manuscripts the angel is shown at work with the 'golden reed' setting out the new Jerusalem, 'four-square', just as surveyors may have pictured themselves using their rods to lay out new towns, some of which, like those in southwest France and east-central Europe, for example, were in the form of 'the holy city', and thus also a microcosm of the wider world that Jerusalem represented (see illus. 15, 22 and 39).

PART III

City-Cosmos Lived

FIVE

Moral Topographies

URBAN EMBODIMENT AND THE BODY POLITIC

Seeing the city as akin to a 'body' is evident not just in Neoplatonic and Aristotelian philosophical tracts of the Middle Ages but in what some individual urban inhabitants were writing about how they perceived their own particular cities and how they thought these ought to be governed. Sometimes such references are explicit, while in other case analogies and links between the body politic and the divine archetype are more implicit. In both, however, there is the sense that to maintain order in the lived city required recourse to the macrocosmic order of things, a view that was supported both through turning to the works of the great philosophers on matters of social or political theory and through turning to Scripture and ideas about Christ's presence in the world. It is as if inhabitants of the earthly medieval city sought to reify their habits and practices by invoking a model of divine rule that governed the world at large, and in the process linking themselves and their cities to the celestial 'city' above.

Much of the evidence for understanding how inhabitants perceived themselves as part of an 'urban body' comes from the later Middle Ages, especially the fourteenth and fifteenth centuries. One interesting earlier example is Lucian's late twelfth-century description of Chester, examined earlier (see chapter One). Looking at this again here it tells us of Lucian's self-perception of his city of Chester as an urban embodiment not only of Christ but the 'body' of the wider world he symbolized. The imagined form that Lucian's description traces out through the orientation and the cross-shape of Chester's streets, the streets 'meeting and crossing themselves', in Lucian's words, suggests that the whole city was seen by him to be imitating the Christian act of marking the body with the sign of Christ's passion, the cross, thus mapping his body onto the 'body' of the city as a whole, and the bodies of those of its inhabitants.[1]

But Lucian also seems to be trying to draw these bodies together by more subtle symbolic means, as follows. He notes the 'positions of the churches' at each cardinal point, and in his description of them starts first with John in the east, then proceeds to turn to the west (Peter the Apostle), then the north (Werburgh the Virgin), and finally to the south (Michael the Archangel).[2] The sequence Lucian maps out, east–west–north–south, is symbolically significant. In his *Enarrationes in Psalmos*, Augustine comments on how 'according to the Greek tongue, Adam himself signifies the whole world. For there are four letters, A, D, A, M, and in Greek speech these are the four letters of the four corners of the earth'.[3] The Sibylline Oracles, a widely circulated collection of early Judeo-Christian prophecies (mixed also with some earlier pagan material), also maintained that the four letters that make up the name 'Adam' correspond with the four cardinal points.[4] It seems therefore that in Lucian's perception of the city of Chester is an embodiment of Adam, and since Adam prefigured Christ an embodiment of *corpus Christi* too.[5] Such an interpretation seems all the more plausible in the context of contemporary images such as *mappaemundi* that show Christ's embodiment of the world (see chapter One) (illus. 36 and 44).[6] Coupled with Lucian's own perceptions of his city as a microcosm of the wider four-quartered, orientated world, as his description makes clear, Lucian's interpretation of Chester (as symbolic of the Body of Christ) is thus a localized reflection of wider medieval Christian understandings of urban embodiment.

Lucian was by no means alone in thinking of his city as a body that symbolized the macrocosm. The model from which it derived – either directly or indirectly – was rooted in the microcosmic analogy between cosmos–city–body that pervaded Christian thought throughout the Middle Ages (see Introduction) (illus. 46). What is significant here is that such perceptions of the embodied city were given local meaning and resonance, connecting a particular individual's city to the wider world around them which Christ embodied. There are other cases of this too, such as a description of Florence written in the later thirteenth century by one of its inhabitants, another clergyman, fra Remigio de' Girolami of the Dominican convent of Santa Maria Novella.[7] In this he remarks on the city's social divisions in tripartite terms, bemoaning how 'the great city is broken into three pieces', with divisions between the Guelfs and Ghibellines, between the 'artisans' and 'the great', and 'between clerics and laymen', to which he concludes 'only our lord Jesus Christ can through his grace unite and combine these divisions'.[8]

Like Lucian, then, Remigio sees the city unifying through Christ's body, the parts embodying a greater whole. It may be suggested that it should be of no surprise that local clergy were drawing such parallels between the urban body and Christ's body – they themselves being a part of both, bodily and spiritually. But there are also examples where civic officials wrote of the urban body and their place in it, drawing connections between themselves as arbiters of local governance and the sovereign authority and divine rule that came from God.

Urban embodiment and urban governance

Towards the end of the Middle Ages a civic consciousness among townsfolk is apparent in what urban officials wrote about their duties and the role of the municipality in governing local affairs. In England by the fourteenth century a growing number of towns were receiving greater degrees of relative autonomy in their government, with citizens gaining the right to elect their own officials rather than have them appointed by a local lord.[9] As part of this process municipalities recorded their civic business such as visits by dignitaries, local by-laws and customs, and in some cases too the means by which their officials were appointed and how their administrations worked. These accounts reveal something of the self-perception of municipal bodies whose governing ruled the everyday activities of its individual members, the bodies of the townsfolk who appointed them. A case in point is that of Bristol, where a late fifteenth-century town clerk, Robert Ricart, set out in a book he called the 'Maire of Bristowe is Kalendar' the origins of the town's legal privileges, rights and customs, and the names of the town's previous mayors, as well as the rites of their investiture.

In the same book that Ricart drew a stylized view of the inner walled city of Bristol, apparently in imitation of the holy city (see chapter One), he explains the process by which a mayor is appointed and the civic rituals that took place when the old mayor was replaced by the new. The mayoral appointment was an annual civic event occurring in September. For all those who constituted the body politic of Bristol – the mayor and the sheriff as well as 'all worshipfull men of the Counceill of Bristow' – the proceedings began with a call by the mayor for divine protection.[10] Ricart records that on 'the morowe upon the daie of the exaltacion of the holy Crosse', on gathering in the council chamber in the Guildhall 'before all the Cominaltee there', that the outgoing mayor 'exhort theym all and euery of them with a pater noster and an ave to pray the Holly Goste to be at their seid election.'[11] Later, after votes were cast and the new mayor chosen, the ceremony for investiture was held: on Michaelmas day and 'byfore all the seide Comyns' the outgoing mayor 'of this honourable Citie' spoke of his service and gave gratitude, and then held 'a boke' in front of the new mayor while the town clerk 'rede the Maires Othe and his charge concerning his office', to 'save and kepe this . . . Toune of Bristow' for the 'oure alther liege lorde' the king.

The Bristol mayoral ceremony – described and depicted in the *Kalendar* – was a symbolic display of civic embodiment, of the city's political constitution; a 'body politic' whose authority came from God, for in the eyes of the mayor and his council, their capacity to rule in Bristol was divinely ordained, a sanctification that was further reinforced by the perception that they were working to the common good and with the king's endorsement. It was thus an affirmation of the city's council as *the* legitimate authority acting for the perceived benefit of the whole urban community and also the reigning monarch.

The investiture process that brought together 'all the Cominaltee' of Bristol in one place, the city's council chamber, unifying the ruling urban body and in their minds embodying the city, was at the same time paraded to the outside world of the city's populace. First by the new mayor having to dress his house 'in as plesaunt and goodly wise as kan be devised' and then pass through the streets of Bristol from his house to the council chamber 'in as solempne and honourable wise as can devise do to his own worshippe [i.e., to his own benefit], and to the honour, laude, and preysyng of alle this worshipfull towne.'[12] Then second, after the mayoral ceremony, in a further procession which involved 'all the hole Counseille', including the old and new mayor, who after dining began to 'walke' from the High Cross situated in the centre of the circular walled city (as depicted by Ricart) to the church of St Michael, and then 'to retorne to the new Maires hous, there to take cakebrede and wyne' (all of which has symbolic resonances with Eucharistic processions, see chapter Six below).[13] Not only, then, does the mayoral ceremony embody the city politically, as a body of governance, it also provided a means of connecting this body politic with Bristol as a spatial 'body', rendering political authority visible to all (not simply to a select few) while assimilating at the same time both the political and the spatial in one unified embodiment that was 'the city'.

Ricart's record of Bristol's corporate constitutional make-up is revealing of its perceived *corporeal* nature. There are similar glimpses of self-perception of urban embodiment from elsewhere too in England. Two particularly telling examples come from Beverley in East Yorkshire, where again the voiced view is that of the town's officials, and hence the body politic. In 1411 'a short conference was held' by 'worshipful men of a worthier sort' to discuss the Corpus Christi play performed 'to the praise and honour of God and of the Body of Christ, and for the peaceful union of the worthier and lesser commons of the town of Beverley'.[14] Here the town is perceived like Christ, as a 'body' through which the town is unified as one.[15] This sense of embodiment is referred to once again by officials in 1498 in a 'new order as to the election of governors', where the contention is forwarded that 'itt is consyderd by the hole body of the sayde towne that the . . . nowmer of 30 persons is to large to be had of the eleccyon for the worschyp and honesty of the sayde towne.'[16] In the fifteenth century the self-perception of Beverley's 'body' was thus one which comprised all the townspeople – 'the hole body' of the town – including 'worthier and lesser commons'; but the language of this embodiment came from the town's local authority, as articulated by its officers who kept its civic orders and statutes and wrote them down.

What we are seeing then, in the case of Bristol and also Beverley, is a perception of urban embodiment fostered through and by the activities of the ruling body (and bodies) within a 'body politic', that is, the mayor and his officers. The body metaphor helped them legitimate their place as intermediaries in the urban social hierarchy, mediating not only between 'all the Cominaltee' of the town and the king – as ruler of the earthly realm – but also between the city and Christ, the supreme universal ruler. In the context of late medieval England the clearest confirmation of this perceived corporeal role of the governing mayor and his ruling body-politic comes from London's *Liber Custumarum* of the early fourteenth century.

The *Liber Custumarum* was one of London's many civic books kept by the city's officials as a record of their statutes and customs. What marks it out as unusual is the inclusion of an extract from *Li*

Livre dou Tresor ('The Book of Treasure'), an encylopaedic treatise written by the Florentine statesman Brunetto Latini while he was in exile in France in the 1260s. Drawing on Classical works as well as contemporary thinking, Latini's *Tresor* covered practical politics with 'suggestions on the process of choosing rulers, the qualities and characteristics they should have, and the negotiations that need to take place between leaders and citizens'.[17] Book Three of the *Livre* in particular dealt with 'the government of cities' and this is the section which appears in London's *Liber Custumarum*. Curiously the *Liber*'s opening extract from the *Livre* is not actually the first chapter of Book Three but the second, 'on rule and its foundation', along with part of the third on 'who should be chosen to rule and govern'.[18]

The *Liber Custumarum* also begins with its own proclamation, that 'he who wishes to make a good choice of a sovereign governor with dignity or with jurisdiction . . . regard the points that follow', after which comes the *Tresor*'s discussion 'of the three pillars upon which the government of cities ought to be supported'.[19] While the London copyist chose to retain the French of the *Tresor*, certain of Latini's words were changed to make it appear more relevant to 'local affairs' for English readers.[20] In particular, in place of *Sire* the word *meire* is used.[21] This linguistic change occurs early on in the *Liber*'s *Tresor* extract, where divine principles of urban governance and rule are discussed and the reciprocal relationship between the city's mayor and the burgesses is outlined. Here Latini draws a connection between the rule of the world and the rule of the city: 'all rule and all rank is given us by the sovereign Father, who in the healthy establishment of things of the world desires the government of cities to be founded on three pillars – justice, reverence and love.'[22] These three pillars uphold the rule of the city, through which a ruler must 'love their subjects . . . for the common good of the city and of all people', while in return 'all others must love their ruler . . . and give him help and counsel in the maintenance of his office'.[23]

To exemplify this functional reciprocity between ruler and ruled, Latini uses the analogy of the city as a body and it is here that the London copyist inserts his own word *meire*: 'the mayor or governor is also, as it were, the head of the citizens, and all persons desire to have a sound head by reason that when the head is afflicted all the members are made ill thereby.'[24] This passage is important as it clearly shows that the body–city metaphor was known to London's municipal officers, and that they were applying it to their own corporate body whose 'head' was the mayor and the citizens its 'members'. Catto argues there is a 'strong indication' that the officer responsible for copying these *Tresor* extracts was Andrew Horn, the city's chamberlain in the 1320s, whose role was to support the mayor and who had also compiled other sources for the municipality's use at this time.[25] How he had acquired a copy of the *Tresor* is a matter of speculation, though it is clear that Horn was making use of Latini's text to suit his own purposes, 'to establish a detailed, practical procedure' governing London, and not simply reproducing it verbatim as an academic exercise.[26]

Thanks to Horn, then, not only did the political ideas of a Florentine statesman become embedded into a practical civic book compiled in London, but in the process he helped mobilize the idea of an embodied city in the municipal lexicon of late-medieval England's most important metropolitan city. This says a number of things about how the principles of urban government were understood in English cities at this time. It shows that the theoretical and practical ideas about political authority and organization circulating through texts such as the *Tresor* were consumed by localized municipal audiences, and made available to them canons of political thought on principles of divine rule and governance deriving from Classical sources such as Aristotle and Cicero. It also shows that there was self-conscious recognition of the corporeal nature of urban governance, and an understanding that the body politic was part of the divine order. In this sense urban officers working in the body politic were (whether they realized it or not) perceiving their role in Calcidian terms, as ruling parts of a functioning whole through which members of the body as a whole will mutually profit, providing the rules are obeyed.

This urban order was a microcosm of the cosmic order, the 'head' of the city's body – the body politic – having a role analogous to that of the 'head' of the 'cosmic body', which usefully justified and legitimized its authority and rule over the city as a whole (see illus. 5). In drawing directly from the *Tresor*, the London copyist thus manages to show not just that he was well educated in theoretical and practical matters of urban governance but that he could use this knowledge to help ensure that whosoever ruled in the city did so with the self-assurance that they were simply carrying out their work according to God's will by replicating the principles of his divine rule.

London, it could be argued, was special among late medieval English cities, and that what we are seeing in the *Liber Custumarum* is perhaps a little civic self-indulgence on the part of its profligate ruling body. Yet the ideas it contains on urban embodiment were not strange to other town and city authorities in England, as noted with Bristol and Beverley, who also saw themselves in these same terms, albeit articulating them in slightly differing ways. No doubt there was interchange between their respective officials on such issues, not least because many were well-travelled merchants whose commercial and property interests were geographically spread. Bristol's mayoral register, compiled by Ricart the town clerk, includes copies of London's laws and customs of around Horn's time that cover matters of electing officials and descriptions of their duties.[27] In Bristol's mayoral speeches recorded by Ricart in the *Kalendar*, the outgoing and incoming mayor's statements are intriguingly similar to those set out by Latini's *Tresor* which appear in the *Liber Custumarum*, raising at least the possibility that Bristolians knew Book Three of the *Tresor* (or some part of it) and had either gained it from their London counterparts or from a copy that they owned.[28]

In late medieval England, then, it is clear that there were those who ruled in cities who saw themselves as a body politic in which the mayor and council acted as its 'head' while the city's inhabitants were the 'members' that made up the body as a whole. Latini's *Tresor* was not the only source that promoted self-perception of the corporeal city. There were plenty of such 'urban ideologues' circulating with similar resonances. Based on contemporary Neoplatonic and Aristotelian microcosmic thinking on the body and the city, they likewise drew analogies between the body politic and the cosmic body and used divine order to justify the basis of ruling urban bodies.

Ruling bodies and divine orders

Latini's *Tresor* was one of many such political works being written in Italy as well as in France and England. They derived from growing acquaintance with Aristotelian political thought, particularly his *Politics*, as well as political treatises by Roman Neoplatonists, such as Cicero and Plutarch, and their derivatives.[29] They also drew on contemporary political praxis and experience, particularly though not exclusively organization of the Italian city-republics.

These later medieval political treatises, works such as the *Tresor*, reveal how the idea of the body politic was understood and defined by those who were describing urban governance as they saw it, combining the practical politics of the cities that they knew with what they had read on rule and government from the philosophers. Their ideologues set out how a ruling body in a city ought to be constituted upon principles of divine law, and how this provided a mandate to issue laws to keep order in the city. It was these very texts that not only show how the city was being made sense of as a 'body politic', but also, for instance in the case of London and Bristol, then being used as a basis of urban governance. Ideas of *divine law* were thus being used and duplicated by the body politic, whose *urban laws* recreated in the earthly city the same order and hierarchy to be found in the macrocosm, the 'city' above. In the process, the practical politics of later medieval cities reflected and reinforced connections between the body politic and the cosmic body, and made the city a living microcosm.

The 'native Italian tradition of moral-political writing' that proliferated during the thirteenth and fourteenth centuries comprised a wide range of authors whose origins and naming shows how closely their work was tied up with the politics of those cities they were familiar with.[30] Even Aquinas, someone conventionally viewed as somewhat detached from everyday urban life, is now seen to have drawn upon his personal experiences of Italian cities and their factions in writing his political treatise, *De Regimine Principum*, of the 1260s.[31] Written ostensibly for informing civil subjects as much as 'the education of princes', Aquinas was using scripture, Roman history and 'the political reflections of Cicero', additionally 'fortified by observations from experience'.[32] In *De Regimine* 'Aquinas' advice was both politic and practical', and in it he 'compared the government of men to God's government of the universe, not because monarchy was of divine origin but because the nature of rule in itself, the pure principle of government, was analogically the same', whether it was 'divine or human'.[33]

Aquinas argues from the principle that 'the ruler cannot create the city out of nothing' and 'so he must look to the government in nature to learn what his duties are in ruling', and this includes looking to God's ordering of the parts that make up the universe and the soul's ordering of the body's 'individual functions'.[34] He had already set out in his *Commentary* on Aristotle's *Politics* how reason 'creates the city as "it orders a multitude of men into *unam quamdam communitatem*"'.[35] This 'multitude is governed by the reason of one person' whose rule governing a city is to derive from 'divine governance', from God's universal laws,[36] through which 'he directs men towards the common good', securing 'a peaceful unity' to 'promote the welfare of the community in such a way that it leads fittingly to the happiness of heaven'.[37] This depends upon a mutual bond between the ruler and the ruled: the three pillars supporting urban government that Latini spoke of in the *Tresor* – justice, reverence and love. Aquinas was writing *De Regimine* as a practical political treatise, to demonstrate to both ruler and ruled their interdependence and collective interest in 'promoting unity and peace in the city', the legal basis of which was to be found in a common divine order that ruled in both the macrocosm and microcosm.[38]

The principles of urban governing which Aquinas was discussing in *De Regimine* were echoed in other similar contemporary works 'in the Italian tradition of practical political literature' of the fourteenth century, including John of Viterbo's *De Regimine Civitatum*, Guido Vernani of Rimini's *Commentary* on the *Politics*, as well as Ptolemy of Lucca's continuation of the *De Regimine* text itself.[39] These works are characterized by a blending of Aristotelian and Ciceronian ideas on the city and its governance. A case in point is Marsilius of Padua's *Defensor pacis* ('The Defender of the Peace'), written in the 1320s and '30s while he was serving 'as an advisor to and delegate on behalf of several important figures in northern Italian politics'.[40] Marsilius frequently cited Aristotle's *Politics*, for example in his discussion of the likeness between the city and the body:

> Let us assume with Aristotle in Book 1, chapter 2, and Book 5, chapter 3 of his *Politics*, that a city is like an animate or animal nature. For just as an animal, well disposed according to its nature, is composed of certain proportioned parts ordered to one another and acting in mutual communication for the good of the whole, so a city is constituted from certain parts, when it is well disposed and instituted according to reason. The relationship therefore between the animal (and its parts) and health, is seen to be similar to that between the city or kingdom (and its parts) and tranquillity. [41]

The city that Marsilius sees is analogous to a body, embodied therefore, but he also pursues this theme further in looking at how the city is itself formed of bodies; bodies of citizens and a ruling body. Again he uses the *Politics* to make his point, explaining a process of governance and law-making by 'the people or corporate body of the citizens, or the more dominant part of them, by means of their election or volition expressed through words in a general assembly of citizens', and arguing also that 'the authority of making or instituting law pertains entirely to the corporate body of citizens or its more dominant part.'[42] This 'corporate body of the citizens' is itself ruled by another body, for 'there ought to be one ruling body in a single city', a body that comprises 'many persons' who act as one.[43] The laws made by this ruling body 'held society together and was necessary for achieving the common good and securing the permanence of government'.[44] Not only then did Marsilius think the city was like a body in terms of its functions and composition of parts; it was also ruled by a body governing by consent of the corporate citizen body as a whole. Such political ideas, argues Nederman, owe as much to Ciceronian as Aristotelian thought.[45]

Canning suggests Marsilius' biological 'view of the body politic' as 'this-worldly and naturalistic' came from his 'medical training'.[46] Yet it could also be said that similar views were widely held among his contemporaries at the time Marsilius was writing, and that their conceptions of the body politic came not just from Aristotelian sources – including, no doubt, the commentaries that were then circulating – but also from the Neoplatonic ideas embedded in Ciceronian works.

In the later Middle Ages, Ciceronian political thought was derived from 'intermediary patristic sources' based upon *De Res Publica* and *De Legibus*, as well as copies made of *De Officiis*, 'among the most widely read and revered texts in the medieval West', from which readers gained 'naturalistic accounts of society and government'.[47] Marsilius himself makes use of 'Cicero's idea of man's transformation into a civilized creature' and (*contra* Aristotle) 'maintains from the start the community and its government must be guided by the whole citizen body'.[48] Cicero discussed this in *De Officiis*, particularly the principle that 'those who take charge of the affairs of government . . . must "care for the welfare of the whole body politic" and not serve "the interests of some one party to betray the rest"'.[49] These principles appear in the parts of *De Regimine* attributed to Aquinas and Ptolemy of Lucca.[50] John of Viterbo also drew a 'connection between the ruler and the common good', and in his discussion of 'The kind of ruler that should be sought for the city and elected as *potesta*' placed an emphasis on the need to have a ruler 'who will rule the city in justice and equity', using Cicero's *De Officiis* to list 'the various qualities a ruler must have'.[51] It is easy to see why such ideas on rule and urban order would have had appeal to those political writers of the fourteenth century drawn from politically riven Italian city-republics such as Padua, Rimini and Lucca.[52]

More broadly, Cicero's political thinking entered the medieval political realm through works such as Latini's *Tresor* and Christine de Pizan's *Book of the Body Politic*, which were being read by political decision-makers not just in Italy but elsewhere too, by the likes of Andrew Horn in London, for example.[53] This blending of Classical sources characterized the political writing of the thirteenth and fourteenth centuries and further disseminated principles of urban governance based upon notions of rule by a body politic. The use of the body metaphor and the principle that 'a well-organized political community will resemble a healthy human body' was also present earlier in twelfth-century practical treatises on governance and rule, such as John of Salisbury's *Policraticus*.[54] It was similarly influential on Neoplatonists such as Alan of Lille and William of Conches, thanks to earlier Boethian and Augustinian renditions of Cicero's work along with Calcidius' commentary on Plato's *Timaeus* (see Introduction). Indeed, these later medieval conceptions of the embodied city bear close resemblance to the Calcidian model of urban governance that was so important in eleventh- and twelfth-century Christian thinking. It likewise proposed that the city is like a body, being a whole made up of parts, 'members', that are ordered hierarchically and which function individually yet together to the mutual benefit of all.

A common philosophical thread not only tied together practical political writing of the later Middle Ages therefore but also extended back to theoretical Classical notions of the 'body politic' ideals of Cicero and Aristotle and the Neoplatonic idea of the embodied city as a microcosm that imitates the hierarchy, functions and order of the macrocosm, reinforcing the divine order of things through obeying universal rules and laws. For those writing on the government of cities in the later Middle Ages, the embodied city was a useful model from which to provide a rational and justified practical basis for urban governance. It was an ordered, hierarchical and functional exemplar, visible in both macro- and microcosm alike, each governed by the same divine laws. The body metaphor connected the city's body politic with the cosmic body as a whole, for 'the hierarchy that prevailed in human affairs was included in and justified by a greater hierarchy that prevailed in the cosmos . . . governed by God, the

Perfect Being.'[55] Urban governance thus derived and extended *from* the cosmic body, whereby the political order and rule in the city imitated that of the macrocosm while what was undertaken in the city by the body politic was simply one part of a broader, divinely ordained hierarchy in which the city's ruler was its 'head'.

No wonder, then, that mayoral investiture ceremonies and rituals of medieval towns and cities, such as the one described in Bristol by Robert Ricart, had overtones of a royal coronation; for the mayor was acting like a king, a parallel by no means lost on those taking part, as is evident in the case of the very elaborate and highly symbolic mayoral ceremony of London, for whose civic edification Andrew Horn had composed his version of the *Tresor*.[56] The city's ruling body – whether *podesta* or *meire* – therefore had a perceived divine basis: it had the power to govern the city using laws modelled on those that governed the universe, and to instill in the city a divine order that mirrored that of the whole cosmos, replicating its embodied form in the city's.

URBAN LAWS AND ORDERING TOWNSPEOPLE

A town or city's legal basis as a corporate body was of course determined by its sovereign. It was through having a charter issued or ratified by a monarch that the urban body politic gained its legal legitimacy to rule over and govern the lives of urban inhabitants. Such charters set out the political and municipal organization of a town or city, giving its rulers the powers to rule. These enshrined a process of political incorporation, legally instituting a town or city as a corporate entity, a unifying and ruling 'body' with its own economic and judicial privileges and rights, including the right to create local urban laws to control what townspeople could and could not do.[57] The charter itself therefore symbolized and legitimated the body politic, while at the same time forging a link between the rule of the civic body, the mayor and his council and their supreme rulers: the superior bodies of the king, as their earthly ruler, and Christ, too, the universal ruler. Through all their bodies his divine rule extended from the cosmos to the city, by which the life of the city and its inhabitants could be ordered in ways that reflected the overall order of the cosmic body. The remainder of this chapter explores this idea further, looking at how the laws issued by ruling urban bodies ordered the city socially and spatially in ways that created hierarchies and boundaries among townspeople which mirrored the perceived functional ordering of the cosmos with its hierarchies and divisions.

Urban laws and their moral topographies

The body metaphor provided ruling bodies with a means of unifying the embodied city while at the same time retaining its internal social differentiation and hierarchies.[58] This can be seen in the way urban laws were used by political authorities to place certain social groups in certain spaces, to order townspeople both inside and outside the urban body, and determine who was where. There are three dimensions to this. The first is that urban laws issued by ruling bodies derived from a moral ordering, a *moral topography*, of the cosmic body at large, where those higher up the hierarchy occupied an inner place in the 'city' while those lower down lay at its margins (see Introduction). The second is that these urban laws marked out hierarchical social differences and boundaries among inhabitants in ways that reflected this moral topography, making the earthly city a living imitation of the cosmic 'city'. The third is that these bodily orderings of townspeople were a means by which marginal social identities were reflected and reinforced, again in an image of cosmic order, to maintain social order and difference in the urban body as a whole.

The ruling body of the city reified its own place as the governing 'head' by perpetuating the universal order of things through its local laws. Their justification for this – if it were sought – was to be found in the political writing of the age, and the principles of divine governance that were seen by Aquinas, Latini, Marsilius and others to be the basis of urban governance, as discussed earlier. A civic corporatism characterized urban governance throughout the Latin West. Spurred on by aristocratic and ecclesiastical overlords keen to expand the wealth of their earthly realms, the inhabitants of towns were provided with a degree of political status and autonomy that favoured both the grantor, the town's overlord, and the townspeople, fostering urban development in the process through trade, property and justice.[59] Self-governance and relative freedom from overlords was based on fixed annual sums raised locally by the urban community. How a particular town or city came to acquire its legal status and local customs is a subject of much debate among historians. By and large, their focus has been on the evolution of political urban bodies as legal entities in their own right, and the constitutional history of

European urbanism has been amply told from this perspective.[60] What has received less attention, however, is how urban laws derived from a body politic relate to a cosmological model of urban governance and order.

'All social categories' in the Middle Ages 'were first and foremost legal categories',[61] and urban laws provided ruling bodies with the means to create social distinctions to differentiate between and favour certain groups of townspeople over others. Laws ordered the citizen body by defining status and privileges among urban inhabitants. Through them a person's place in the social hierarchy (high–low) was linked with their dwelling-place in the city (inner–outer). This mutually reinforced social and spatial ordering of townspeople was important because through it the inhabitants of a city were *placed* in the eyes of themselves and others, defining and projecting their identity, a *placed identity*. Hence some individuals and groups were better placed, socially and spatially, while others were being ruled against and marginalized by those who were issuing laws, the ruling lords, kings or municipalities: the body politic which controlled the everyday lives of townspeople.

Urban laws were used to differentiate those on the 'inside' from those 'outside', creating a moralized placing in the city where the lowest social orders were located at the urban margins, 'the city's outskirts', 'the lower portions of the body' in Calcidian terms.[62] With its functional and hierarchical distinctions between different social 'orders', the city's 'moral topography' mirrored that of the city-cosmos, the macrocosm (see Introduction). Urban laws therefore provided a further link between the city and the cosmic body, one forged by the body politic: just as God issued divine laws which governed all, the ruling head of the city meted out its urban laws; and just as the universal ruling body ordered the celestial 'city' by divine law, so the earthly city's governing body used its laws to order the urban social body in ways that imitated the 'moral topography' of the cosmos and the heavenly city (illus. 5, 36 and 38). What follows examines how ruling bodies used urban laws to order townspeople's bodies according to these divine orders. To do this the first part of the discussion looks at the application of urban laws in England by Norman overlords. It then turns to the laws issued by municipalities later on in the Middle Ages.

Written charters make it clear that ruling bodies governing inhabitants used laws to define social and spatial boundaries in the urban body as a whole, creating a moral bodily ordering of the city.

To create chartered towns overlords in the twelfth century used 'families' or models of urban law. One example is the Law of Breteuil. The customs of this small town in Normandy provided a legal blueprint adopted by Norman lords in England and Wales as they sought to enfranchise both new and existing towns with borough status in the eleventh and twelfth centuries (illus. 56).[63] The practice they adopted, borrowing legal privileges of an existing town and applying them to another, was not confined to Norman Britain. It was commonplace in other parts of Europe too, where lords were likewise trying to encourage urban growth and stimulate people to take residence in new towns, especially in those regions newly acquired through territorial expansion.[64] In east-central Europe, for example, a family of urban laws deriving from the older-established towns of Magdeburg and Lübeck were spread across the area between the Oder and Elbe rivers in a process of colonization under Germanic lords during the twelfth and thirteenth centuries.

In those regions of Europe where lords were seeking to consolidate hold over newly acquired lands and people, as the Normans were in England and Wales, there were clear advantages in issuing urban laws based upon the customs of their home towns. For not only did urban laws help to populate an area through urbanizing it, the laws themselves could be made to be more favourable to incomers arriving from elsewhere rather than the local existing people: they could be used to differentiate urban bodies.

At the time of the Norman Conquest in 1066 the kingdom of England was well served by an urban network, with towns and cities that had gained their own customs and laws from the crown at various times in the past.[65] After 1066, as English fiefs were handed out to the men who had supported William the Conqueror's victory, charters were issued to established towns and cities, as well as new urban foundations, benefiting some of the townspeople with social and economic privileges, but not all. Whereas the existing English inhabitants kept their old laws, as they had before 1066, newcomers in the towns were given fresh laws granted by the new ruling Norman ascendancy. At Shrewsbury this disparity caused English burgesses to complain to Domesday commissioners, saying they were placed at an unfair disadvantage because of the new customs.[66] Up and down the land, the Law of Breteuil provided a model for these new laws, attracting newcomers from outside to come and live in the English towns because they offered favourable terms.

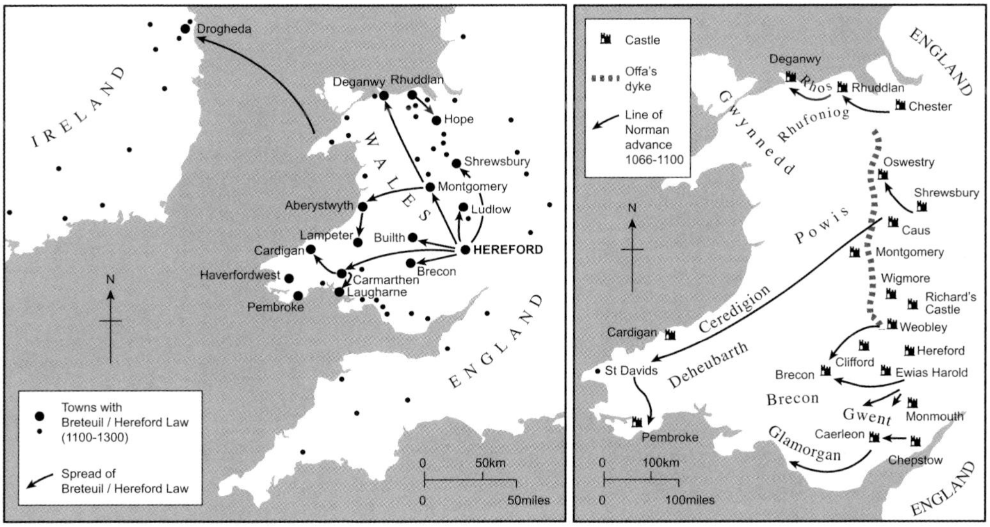

56 The spread of the Law of Breteuil from Hereford, in Anglo-Norman England and Wales.

This policy of watering down English towns by encouraging 'outsiders' was surely a deliberate one, for the towns were often the most troublesome centres of English resistance for the Norman overlords, as was the case with Exeter, Norwich and York.[67] With the new French, Breton, Flemish and Norman settler-migrants, the Englishness of England's old established urban centres was eroded, to the political and commercial profit of the new Norman overlords. But there was more to this policy than simply promoting immigration and urban growth. The laws the Normans were introducing also had the capacity to marginalize English townspeople, for they marked out the town's inhabitants both socially and spatially: creating and reinforcing social boundaries by placing newcomers on the 'inside', and hence higher up the social hierarchy, and placing existing townspeople on the 'outside' and thus lower down on the urban margins.

Hereford was one of the first English towns to have been issued with the Norman Law of Breteuil (illus. 57). The original charter for Hereford does not survive but Bateson managed to reconstruct it from fragments of later borough customs that had been based upon Hereford's.[68] Breteuil law was first granted to Hereford by William FitzOsbern very soon after the Norman Conquest when he became Earl of Hereford and was given the marcher lordship by the king, William I.[69] It did not apply to everyone in the town however, but only to those whom contemporaries labelled 'French' (*Franci*), coming to Hereford to take up residence. Local English townspeople were

57 The early Norman expansion of Hereford (England).

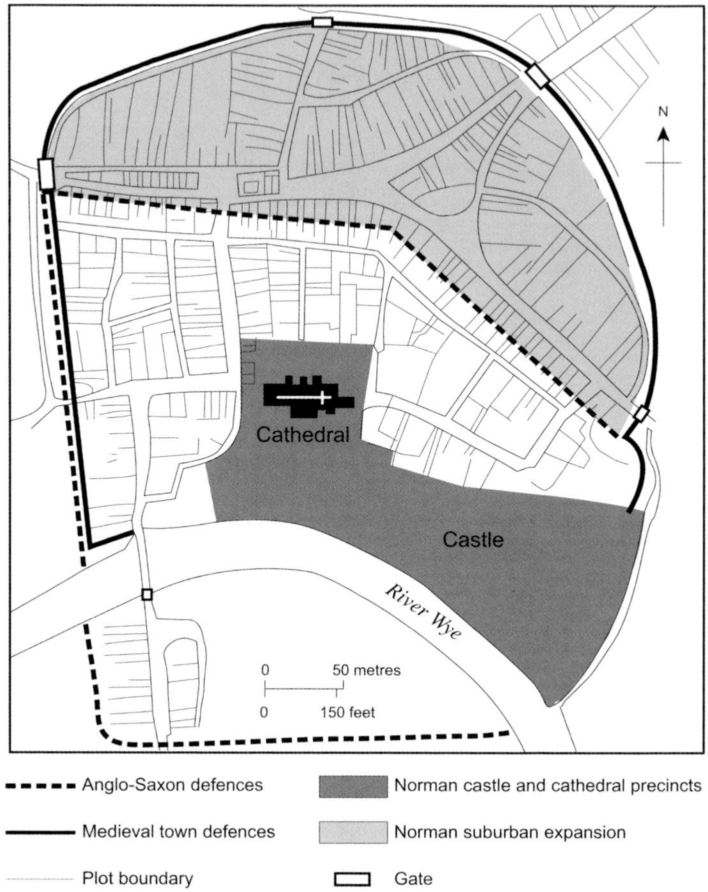

excluded by the new urban customs issued by the earl. In 1086 this social division appears to have been a real one, for Hereford's entry in the Domesday Book distinguishes between English burgesses, living according to their old customs, and French burgesses, who were paying a maximum amercement of twelve pence annually.[70] This constitutional difference put the newcomers on a much more favourable financial footing than their English counterparts. Bateson suggests that there were in effect two boroughs in existence at Hereford at this time – one English and one 'French' – dividing the town's population.[71] The new French borough appears to have been sited just outside and adjacent to the defences of the older established English borough, arranged around a large triangular marketplace with spacious building plots fronting it, and enclosed by another circuit of defences (see illus. 57).[72] Here was the constitutional and topographical boundary

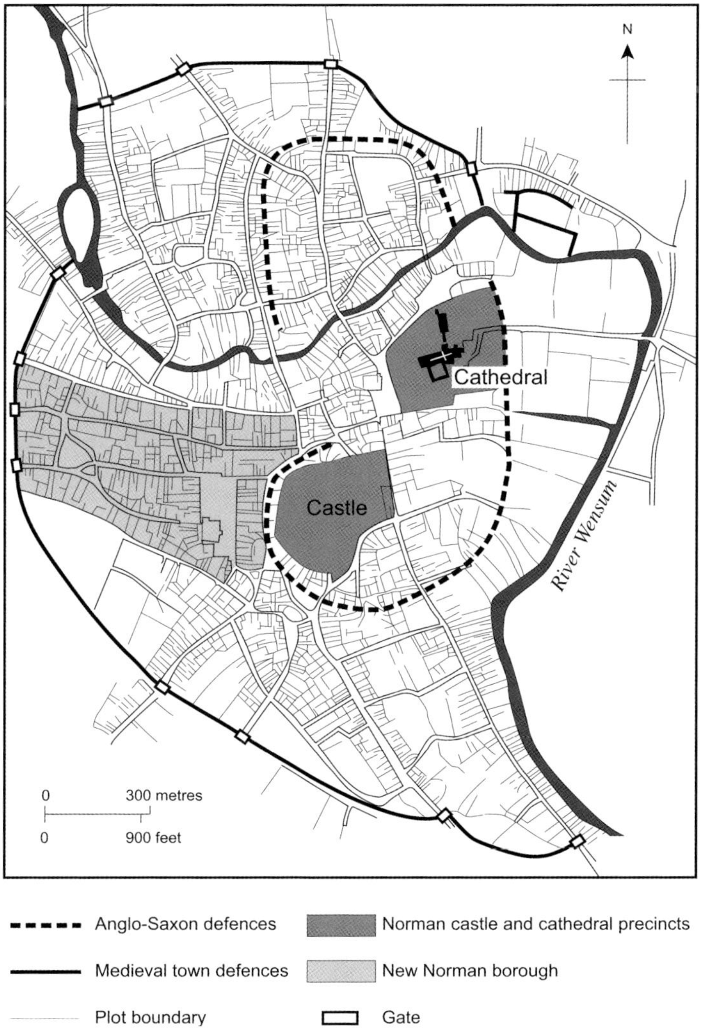

58 Topographical changes in Norwich (England) following the Norman Conquest.

between those in Hereford living according to French and English law.

The boundary thus differentiated between two social orders, between those included under the new laws and those excluded by them; one was superior, the other inferior, both morally placed by the ruling body. Elsewhere, separate yet proximate Norman boroughs were created by Norman overlords. At Nottingham, immediately following the Conquest, a new borough was established adjacent to the English borough, while in Norwich (illus. 58), Bristol (illus. 59) and Southampton, topographical and toponymical evidence reveals that

'French' towns were similarly placed alongside existing 'English' ones.[73]

Marginalization and placed identities

Urban laws, then, were used to create social boundaries by legally and spatially defining certain groups of people, placing them in certain areas of a town. The spaces these groups inhabited were the places in the urban body as a whole where they were perceived to belong. The Normans were not alone in using urban laws this way, for they were being used to instill social boundaries all over the Latin West.[74] It is just that in the political and cultural context of territorial expansion, when success depended on securing control of people and land, that the use of urban space to exclude is especially critical, as it was for the Normans colonizing England and Wales in the later eleventh century. Two centuries later, again as Wales was being further colonized and brought yet further under English control, borough charters issued by Edward I similarly excluded the resident Welsh from living under the same urban laws as English burgesses who moved to take up residence in newly founded towns such as Denbigh and Ruthin.[75] Again, this was a legal separation, defining townspeople's place both socially and spatially, with the Welsh inhabitants having to live outside the English borough.

It is this mutual spatial and social exclusion that is of particular interest, for in all these examples it shows an attempt to use urban laws as a means of reflecting and reinforcing placed identities, of defining social hierarchies and orders by spatial means. Here we can begin to see how by placing certain people in particular locales in a town or city, urban laws helped to forge a placed-identity reflecting the macrocosm's moral topography with its hierarchical distinctions between centre and edge and inside and outside.

With those inhabitants of Hereford living under the Breteuil customs issued by William FitzOsbern there was something symbolic about where they were placed in the town. The English burgesses remained where they had been, living within the walls of the *burh* that dated back to the time of Offa, but the 'French', as we have seen, dwelt outside the walls of the *burh*, in a new area of the town (see illus. 57).[76] On the surface it would seem that the latter were more marginally placed than the former, for the English occupied the old centre of Hereford. However, the more favourable laws of the 'French' borough, together with the sizeable market area that was created for it, suggests that what was happening was the reverse: that the idea in FitzOsbern's mind was to *marginalize* the English burgesses, legally, socially and spatially, by recentring the town onto a new focus, a new 'centre', the French borough.[77] What had been the edge was now the new centre of Hereford, and those who lived in it and enjoyed its laws were elevated socially and economically in the town as a whole. This was a social reordering of Hereford; by placing those bodies deemed to be marginal – the English burgesses – at the edge, and re-placing the centre and those within it, English burgesses were being spatially marginalized to reinforce their marginal place-identity. So not only were they legally less privileged, by having to live under their less favourable customs, they also – through the place they inhabited – came to be *seen* to be more inferior. Underpinning this social reordering of the townspeople was a belief in the moral superiority of being at the centre and a reciprocal inferiority by being at the edge, at the margins: a moral topography of urban placing.

The process by which this new ordering came about is little known, but the situation at Hereford was matched in other English towns by similar reversals of centre and edge, core and periphery. Under the Normans the same changes occurred, for example, in the early years after the Conquest in Norwich, where a new borough, based again on a large marketplace, recentred the commercial focus of the town away from the earlier Anglo-Saxon core, a process of marginalization further compounded by demolition of English areas of housing to make way for the Norman castle as well as an expanded cathedral precinct (see illus. 58).[78] The 'new' Norman marketplace remains the centre of Norwich to this day. At Nottingham and Bristol, too, areas of expansion under Norman lordship were attempts to refocus the urban core and with it marginalize the English urban communities by effectively placing them on the edge. Again at Nottingham this was quite a success, and in this case the English and French boroughs continued as separate entities for centuries afterwards, but this was not so in Bristol, where the Anglo-Saxon core retained its centrality and the new Norman market area declined such that by the thirteenth century it had become known as the 'old market' (illus. 59).[79]

In all these cases it is the spatial manipulation of the social order of the town that helped define 'inner' and 'outer', placing some people at the centre and others at the edge. This moral topography that the Norman lords were creating in their newly acquired towns

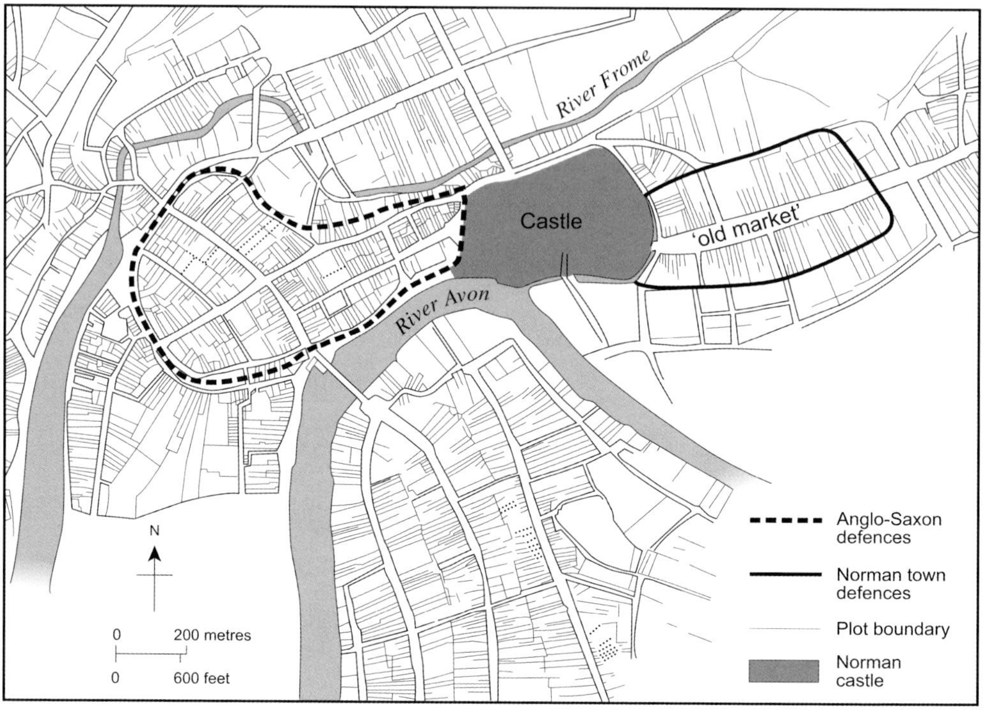

59 The early Norman development of Bristol (England).

was itself surely deliberate, for it succeeded in making the English towns less of a nuisance for them in the early years after the Conquest. The social and spatial divisions created between English and French townspeople were mapped onto them bodily, placing them and marking them out as different, 'othering' them. How far their lords – the ruling bodies – were consciously being influenced by ideas deriving from a model of the moral order of the imagined city-cosmos is almost impossible to say for this period, though such Neoplatonic ideas were circulating in both France and England at this time among Christian thinkers. Nevertheless, for their successors, the municipalities and urban governments of the later Middle Ages whose task was to control and order the corporate body of citizens, the body metaphor and its symbolism was widely understood, as we have seen.

Exclusionary practice and the body politic

Writing of the city being governed in imitation of the model of the government of the universe, as Brunetto Latini and others were doing in the thirteenth and fourteenth centuries, it is evident that the analogies between the two bodies were appreciated by those who were themselves members of an urban body.[80] Civic governments such as ruling overlords used urban laws to instill social order in the city, differentiating townspeople and reinforcing internal hierarchies.

Through issuing laws the body politic was the decision-making head of the corporate city, and the laws it made regulated the lives of urban inhabitants, drawing boundaries between them both socially and spatially in ways that imitated the moral ordering of the archetypal 'celestial city' as described by Calcidius and depicted in images of the heavenly Jerusalem and the wider universe (illus. 5 and 38). To this end, the divine laws that justified and legitimated the rule of the body politic were diffused by it throughout the whole body of the city by regulating where certain social groups were placed (as had been the case with the urban laws issued by Norman overlords).[81] Again, the most marginal of the groups against which the body politic legislated were made to occupy the most marginal locations in the city. One particularly clear indication of this is the handling and placing of perceived social 'deviants' whose bodies were expelled and marked by the actions of the ruling urban body to keep them in their place and make it clear to townsfolk that not only was their city ordered by the body politic 'for the good of the whole', but that this social and spatial ordering derived from divine laws governing the city itself.

Of all the social groups of the medieval city, lepers and prostitutes were considered most deviant and across the Latin West civic governments legislated against them to put them in their place. One particularly well-documented case where this occurred is Bristol, described in the twelfth century by the author of the *Gesta Stephani* as 'amongst the richest city of all in the country, receiving merchandise by sailing-ships from lands far and near'.[82] A charter made Bristol a city and county in its own right in 1373, and set down for the first time the organization of the council and its members, though the origins of Bristol as a *burh* go back to at least the tenth century (illus. 59).[83] It was not until the twelfth century that charters of urban privileges begin to appear, such as one granted by the king in 1155, and a later confirmation of 1188 set out by John Lackland, Count of Mortain, then holder of the town's castle.[84] A mayor had existed certainly by the early thirteenth century, and although the mechanism of urban government operating at that time is not known in its details it is clear that there was a council elected by the *communitas*.[85] This constitutional history of the city was that which Robert Ricart was setting out in the fifteenth century in his *Kalendar* with its descriptions of mayoral investiture that seem to bear such close correspondence to those outlined beforehand by Horn for London using Latini's *Tresor* (see above).

As part of the ongoing development of Bristol's civic government during the fourteenth century, certain proclamations and ordinances were made and it is these that reveal the attitudes of the city's body politic – the ruling body – to marginal groups living within its sphere. One such record is the *Little Red Book*, begun in 1344 by William Colford, the town's recorder, in which were kept 'the ordinances, customs and liberties of the Commonalty, and other "necessary things to be inviolably kept in perpetual remembrance", for . . . many good customs had been abused, and some almost forgotten.'[86] Sometime late on in the fourteenth century, under the heading *Proclamationes Ville Bristollie*, a subsequent recorder of the *Little Red Book* wrote out 'a series of forty regulations for trade and the welfare of the town generally'.[87] Two rules in particular concerned Bristol's lepers and prostitutes: one declared that 'no lepers stay in the town under pain of imprisonment', and the other 'that no common woman wander about the town without a striped hood'.[88] In being ruled against, these two marginal groups were marked out bodily by the body politic – the lepers by their exclusion from the town and the prostitutes by their clothing – both defined therefore as 'other' by the corporate city as a whole.

The bodily exclusion of Bristol's marginal groups was further affirmed in a charter granted to its burgesses in 1331 by King Edward III. This gives an even more revealing indication of local political attitudes towards them. The charter states 'it is ordained and agreed that in future no leper reside within the precincts of the town, nor any common woman within the walls.'[89] This treating of both groups under one regulation is itself telling, tarring them with the same brush as it were, but more significant is the action to be taken against them: expulsion from the walled area of the town, casting them out to the edge, to the urban outskirts. The suburbs were seen to be the most suitable place for such perceived deviants, reflecting and reinforcing the idea that the outsider in society was always at the physical margins, placed away from the ruling body, the city's head, lest it be corrupted to the detriment of the whole urban body, the citizenry who made up the corporate city at large.

Bristol's lepers and prostitutes were thus excluded from the town by the town, the corporate body having 'agreed' to this. They were placed socially and spatially beyond the inner core and displaced to the suburbs, placed on the outside. The walls circling the town symbolically marked out and delimited this social and spatial boundary, separating out the two, urban and suburban. It was a boundary in a moral sense between inside and outside, defining a moral hierarchy and ordering the constituent members that made up the urban body. As an image of the world, Bristol's moral topography thus paralleled that of the 'celestial city'. It is evident both in its idealized circular form depicted in the mayoral *Kalendar* by Robert Ricart (see illus. 42) and by the way the body politic sought to regulate the lives of the city's inhabitants: the local urban laws created an order within Bristol through which its body corresponded to the 'body' of the cosmos governed by God's rule. As far as the laws concerning lepers were concerned there was a clear divine basis, as it was rule decreed by Scripture, as Leviticus instructs: 'all the days wherein the plague shall be in him he shall be defiled; he is unclean: he shall dwell alone; without the camp shall his habitation be'.[90] By being removed spatially from 'the camp' (cast out) the leper becomes an outcast, removed from society. Moreover, it is not just a bodily ejection, for Leviticus also states that if their home is seen to be likewise infected with the disease then it too is to be cast out: 'if the plague be spread in the house', the priest 'shall break down the house, the stones of it, and the timber thereof, and all the mortar of the house; and he shall carry them forth out of the city to an unclean place'.[91]

The liminal spaces of the city's outskirts are thus seen to be the appropriate place for what is deemed to be 'unclean'. To the inhabitants of Bristol this not only included lepers (in England by the fourteenth century leprosy was a rare disease) but also prostitutes, and there is just perhaps a little resonance of these words of Leviticus in Edward III's charter of liberties granted to 'the men of Bristol' concerning what would happen to prostitutes inside the walls: that 'if such women be found living there . . . then the doors and windows of their houses be taken down and carried away by the Serjeants of the Mayor (*servientes maioris*) to the house of the Constable of the peace of the same ward and kept there until such women are entirely removed.'[92] Dismantling houses was just one more symbolic act performed by the urban body to displace undesirable elements and socially exclude them.

According to divine law the unclean were therefore legislated against through the 'head' of the urban body, and made to occupy marginal places around the town. The association between deviance and liminality was reinforced by the laws laid down by the ruling body politic acting on behalf of the corporate town, and by the marginal places that supposed deviant activities – sex outside marriage, for example – were clandestinely conducted. The rules enshrined in the town's charters and statutes did not prevent these activities from taking place, but rather the laws simply pushed them to the more marginal areas of the town. This is indicated by particular street and place-names referring to the sexual acts carried out there. In Bristol, a narrow lane, called Love Lane in fourteenth-century records, led from the quayside to Marsh Street, outside the town's earliest inner circuit of walls, close to the port (illus. 59). More explicit is 'Gropelane', first mentioned in 1281, and probably the same as 'Gropecountelane' that lay close to the town wall in 1339, a street that appears to have been situated again close to the quayside.[93] There was also 'Fockyngrove' situated on the boundary of the city's jurisdiction, and recorded in the charter of 1373 that granted Bristol the status of a county in its own right.[94]

Such places were on the margins, whether alleyways or open ground, where norms and laws could be and were transgressed. Yet they reinforced, at the same time, the social and spatial connection between deviance and marginality, reflecting the same moral order that the laws instituted by the body politic sought to enshrine. The Bristol evidence is matched by examples from elsewhere, for instance a Grope Lane at Southampton connected with the marketplace in English Street, while at Norwich another 'Gropecuntelane' lay close to the castle ditch between the marketplaces at Tombland and Mancroft.[95] This close spatial connection between the places where traders and prostitutes both gather has not gone unnoticed by historians,[96] but its broader significance in terms of how such places were projecting and reinforcing an ordered and hierarchical *moral topography*, is something that has been overlooked. The same is also true for discussions of 'leper hospitals', and their locations on the urban periphery, beyond the walls,[97] for here again spatial liminality not only reflected the perceived place of the leper in (or rather outside of) medieval society, but reflected a deeper, divinely ordained ordering of society and space, in which the cosmic body itself was mapped onto and through the living urban body. To this end, the body of the medieval city was a lived microcosm, functionally and hierarchically ordered and orchestrated by a ruling body politic whose instruction – it was believed – came from God.

SIX

Performing Bodies

CORPUS CHRISTI AND URBAN EMBODIMENT

The embodied city was not simply a political metaphor; it was also one that was performed through specific displays of devotion in towns and cities throughout the Latin West. On certain dates in the Christian calendar, and celebrating and commemorating the events of Christ's life, various processions, pageants and plays were performed in streets, marketplaces, open spaces and buildings. Such ritual bodily performances in and through the medieval urban landscape were important for generating and projecting a sense of urban identity and community, the urban 'body' acting as one to honour both Christ and his saints.[1] The performances were thus both individual and corporate, and through passing spatially in and around the city they connected the urban 'body' with the Body of Christ, and hence the city with wider world. Commemoration and celebration of this kind had the 'body' of the city performing as a whole, as well as its individual parts, or members, the citizenry that made up the body as a whole.

In the medieval city, these ritual performances traced out both a sacral history of the world and its cosmological form, encompassing both literally and metaphorically the 'body' of the city, socially and spatially, and bringing together as one the earthly city and the divine Body of Christ. This urban embodiment is explored here first by looking at the cosmological symbolism of the shapes traced out by urban bodies as they performed their rituals celebrating Christ's body along particular routes and in specific locales; second, the cosmogenic significance of the content of the rituals is considered, in particular Corpus Christi plays, which through their performance in the city connected urban inhabitants – the social body – with cosmic history as embodied by Christ. By unifying their own bodies with the Body of Christ, townsfolk associated the form and formation of the wider world with their own urban worlds.

Community and communion

Rituals are performed in order to commemorate some past mythical event, imitating in the present that which has gone before. Religious festivals are the embodiment of such ritual performances, a 'participation in the sacred' that 'enables man [*sic*] to live in the presence of the gods', as Eliade put it.[2] In the Christian tradition, this is most clearly seen at Easter when Christ's entry into Jerusalem and his journey to the cross are re-enacted annually by repeating the procession during Holy Week.[3] It seems that 'for religious man, a reactualization of the same mythical events constitutes his greatest hope; for with each reactualization he again has the opportunity to transfigure his existence, to make it like its divine model.'[4]

Processions have indeed long been integral to Christian faith and devotion. In early Christian Jerusalem, Rome and Constantinople, particular sites of worship were linked through processions, a 'stational liturgy' that Baldovin explains had its roots in Classical Greek and Roman traditions as well as Judeo-Christian.[5] 'The liturgical procession' of the early church 'enabled not only the participation of many people and an expression of their piety, but also the expression of the religious faith of Christians as the "common-sense" foundation of the life of the city'.[6] The urban nature of such 'participatory processions' is of note, as well as the forms that they took and the meanings they had. Processions were either 'supplicatory', 'whose main aim was to beg for God's mercy', or 'mimetic', 'historical repetitions of a famed action', where the city re-enacted events such as Christ's Passion.[7] Both forms were an enduring 'means of expressing the public and cultural nature of Christianity, and an expression of that faith as pilgrimage and process', and both also encompassed the living urban body, such that (in the words of St John Chrysostom) 'the whole city has become a church for us'.[8]

The symbolic significance of a ritual performance undertaken by the city as a collective body was examined some twenty years ago by Mervyn James.[9] In his seminal article on ritual, drama and the social body, he argued 'that the concept of body provided urban societies with a mythology and ritual in terms of which the opposites of social wholeness and social differentiation could be both affirmed, and also brought into creative tension, one with the other.'[10] To demonstrate this he looked in particular at events celebrating Corpus Christi day, especially the processions, pageants and plays to be found in later medieval English towns and cities, such as York, Coventry, Chester and Norwich. The feast of Corpus Christi was itself quite a latecomer in the medieval townsperson's ritual year, gaining popularity across Europe during the century or so after it was authorized by Urban IV in 1264 'to give thanks for man's salvation through God's decision to become man'.[11]

The civic ritual performances associated with Corpus Christi are considered specifically later in this chapter. Of relevance here are James's arguments concerning how, by understanding itself as a social 'body', the city was 'performed' through municipally organized urban processions. He contextualizes the Corpus Christi civic celebrations by drawing upon the 'medieval social background against which the cult was practiced',[12] asserting in particular that at the time the '"body" was the pre-eminent symbol in terms of which society was conceived', and that 'it was the idea of the social order as body which had the widest connotation, and which was the most obsessive and fruitful.'[13] These analogies of city and body are, of course, those scrutinized earlier.[14] What James suggested is that this same concept was physically manifested in and through the urban landscape by the 'body' of the city performing Christ's body. Thus, 'the Corpus Christi procession involved the application of this theme to a specific community and place, presenting in visual form the structure of social differentiation taken up into the social wholeness which was the town itself.'[15]

In some ways, because of its very nature, *corpus Christi* – the Body of Christ – provides the most apt symbol for the embodied city, and to this end James is able to establish a convincing symbolic connection between them. Corpus Christi was after all a feast day created to give the Eucharist greater prominence in the ritual year; the time chosen for it was the fifth day after Trinity Sunday rather than the solemn but more appropriate day, Maundy Thursday, at which Christ told his disciples, 'eat, this is my body'.[16] A sermon, originally written by St John Chrysostom but 'read at matins during the Corpus Christi season' (usually late May to mid-June, depending on when Easter fell), explained how 'Christ hath mingled himself with us, and has infused his body with our bodies, that we may be one together, as limbs of one body.'[17] Considering the contemporary understanding, either through political theory or practice, of the social and symbolic meaning of the 'body', it would be surprising if the municipality – the corporate body – of a town or city had ignored the feast day as a means of reinforcing its authority and identity.

As noted earlier, in towns and cities in England in the 1300s and 1400s there were those in urban politics who saw themselves (as officials in local government) as part of a metaphorical 'body' comprising 'head' and 'members', which, along with the corporal imagery used at Corpus Christi, gave this conception of the body politic particular symbolic and mystical resonance.[18] There were other ritual occasions on which Christ's body was celebrated and commemorated by performances within and by the city, most obviously during Holy Week, but also royal entries into a city, when the urban body welcomed the sovereign in imitation of Christ's arrival in Jerusalem.[19] As ritual performances by which the urban body and Christ's body were symbolically conjoined, the Corpus Christi processions and pageants of the later Middle Ages have attracted most attention, however; in part through the work of historians of theatre and drama interested especially in the plays that developed alongside and associated with civic celebrations, and in part by historians of medieval religion and culture, whose interests are more in the cult of *corpus Christi* and its place in the civic, political and social life of towns and cities.[20]

The procession was fundamental to this embodiment of the urban community and its communion with Christ's body. As James put it: 'the procession, which was itself the wholeness of the urban social body, gathered in unity and concord to venerate the Corpus Christi, itself a central symbol of social wholeness, and joined in this by the massed crowds through which it moved.'[21] The body metaphor was therefore capable of being used by the body politic to maintain both social unity *and* division, as noted earlier (see chapter Five), and 'feasts, processions and dramatic spectacles provided the town with a language and grammar, as it were, to express this union of differences.'[22] By their very participation in the festivities on the day – not simply the 'dramatic' pageant but also the 'liturgical' procession – the audience itself was a bodily expression of the local urban populace: the *communitas*, the city as a people, were gathering on Corpus Christi day to collectively take the sacrament and thus be in communion with Christ. Those present, whoever they were and whatever their role, not just those urban elites who found in the idea of the body a divine justification for their place in the city,[23] or the performers in the pageants or processions,[24] were all encompassed by the performance and ritual, circumscribed by the unifying Body of Christ, at one with him. In this sense, the body was the city and the city became an embodiment of Christ.

The debate opened up by James's article on Corpus Christi and civic ritual has prompted a more nuanced interpretation of the 'body' metaphor and the contested meanings it had in its use in urban contexts during the later Middle Ages. Yet two related aspects relevant to the symbolism of the urban body remain under-explored: the first concerns the spatial dimensions of performed civic rituals, while the second is their cosmological symbolism. Their significance in the ritual performance of Christ's body by the civic body is alluded to by James but not developed as part of his argument. Since then others have begun to recognize this but still to no great extent. For although Beckwith, for example, noted that space 'is too simplistically conceived by Mervyn James as the *tabula rasa* of the procession', and that it 'comes to have meaning through practice itself', she does not explore how.[25] Similarly Rubin recognizes that in constructing and conveying cultural meanings space matters, as others have, but in the same way that they have criticized James for not having a nuanced understanding of the 'social body' and ritual they too may be accused likewise with their discussion of space. For a start, it may be worth distinguishing *how* urban spaces have meanings through the practices of those who negotiate and inhabit them, and what role ritual performances had in this regard.

Performances in particular places, for example, attach symbolic significance to particular parts of the city. Then there are the spatial patterns and forms traced out by the performances themselves, especially the routes of processions, and the significance that these had in the minds of their participants. And of course, the body is a spatial as well as a social entity, and hence Corpus Christi is a spatial as well as a social metaphor for the urban 'body'. These, and other 'spaces' of bodily performance, require consideration. To do so will also begin to address the second of the two deficiencies identified above; the cosmological symbolism of ritual performance. Here again, James hints at this in the idea of the embodied city,[26] but does not make the case himself, and while Rubin notes how, 'in the medieval context, the body was a metaphor of metaphors, embedded in the sacramental cosmology as well as in personal experience', the cosmological symbolism of the Corpus Christi processions remains under-explored beyond identifying that through the body metaphor the order of the natural and social worlds corresponded with each other.[27] Those who have come somewhat closer are Nijsten and Boogaart, both writing about ritual performance in Flemish cities.[28] Seeing ritual performances of the urban body as a spatial as well as a social expression of the wider world reveals how the city was lived as a microcosm: for performing Christ's body mapped onto the city the world that he Himself embodied.

THE CITY AS A 'LIVING' WORLD

Since Christ embodied both the earthly city and the heavenly 'city', the processions held on the feast of *corpus Christi* were a *lived* microcosmicism – performed by and for the urban community – conjoining through them Christ's body and the cosmic body. In this part of the chapter this idea is pursued further by examining, first, the cosmological significance of the urban locales and routes of ritual performances, especially those associated with the feast of Corpus Christi and processions celebrating Christ's body, and second, the way mythical times and spaces were performed through the urban landscape by the city, in particular the cosmogenic symbolism portrayed by pageants, plays and processions. While the former is concerned with tracing out the geographies of ritual performances and looking at their cosmological resonances, the latter is concerned with how through these rituals the city enacted Christian conceptions of cosmogony. In both respects the figure of Christ is crucial, for it is through his body that the cosmic body is brought into dialogue with the body of the city. Connecting them all is a social and spatial ordering of performing bodies through which the city becomes a living microcosm of the wider world, replicating its form in the city's performing bodies.

Geographies of performance

In *mappaemundi* such as the Ebstorf map, Christ is depicted embodying the world and its history. His outstretched arms touch the very edge of the world, encompassing all, while the four cardinal points found in his cruciform figure orientate it from the centre of the world, the *axis mundi* that was Jerusalem, the centre of his body (illus. 4). The world was synonymous with his body, therefore, in the Christian imagination (see chapter One). Coupled with this were contemporary ideas of the city and Christ being unified as one through civic rituals that celebrated and commemorated his body; this spatialized Body of Christ mapped onto the city an image of the world he represented, so through performing Christ's body, the city – and its constituent performing bodies – could connect their world with his. They did this through tracing out the form of the world in the spatial patterns of rituals in and through the city, through their *geographies of performance.*

Apart from in some antiquarian studies, relatively few have written about routes and locales of medieval civic rituals in general and Eucharistic processions in particular.[29] The tendency is instead to focus on the narrative content and meanings of performances rather than where they were taking place and what significance these places may have had for participants.[30] Urban space was not neutral or inert.[31] It carries and constructs social and cultural meanings, so it seems appropriate to consider where processions went in the city: which streets or lanes they followed, which marketplaces and churches they stopped at, and so on, for these were not chosen without thought, and in themselves offer insight into the symbolism of the processions themselves. As Ashley has pointed out, these 'processional routes are the clearest maps to the significant power structures within a community, since they are always deliberately designed with references to places that are important.'[32]

It is perhaps surprising then that the performed geographies of processions and their meanings have been so overlooked. To begin to redress this, the following discussion focuses principally on Eucharistic processions and examines the routes they took, and suggests how their forms may be interpreted as ways by which the urban body connected itself with the cosmic body. Hidden in these processional geographies are clues that point to the cosmological significance of the processions themselves. The first task is to expose these geographies, and then to reflect on their meanings, but even mapping out medieval urban processional routes is problematic, for they were written down rarely or described loosely, and in some cases changed over time. With these caveats in mind the following is an attempt to begin to undertake this mapping exercise drawing upon examples of processions for which evidence is reasonably clear. The patterns the routes trace will be seen to conform to those traced out by the figure of Christ in depictions showing him embodying the world, encompassing the city by defining centre and edge, and creating linear axes that divide yet spatially unify the body of the city. Of course, in celebrating Corpus Christi the routes taken by processions are not simply passages to publicly display the civic body, but are the routes through which the Body of Christ is carried and hence unified with the embodied city socially *and* spatially.

The procession of the Holy Blood at Bruges is a Eucharistic civic ritual that has been carefully and innovatively examined by Boogaart both in terms of its geography and its symbolism. He identifies a 'symbolic parallelism and reciprocity between community and cosmos, city and hinterland, head to body, guilds to government, and parts to whole' in the annual ritual held in Bruges involving the city's veneration of a phial containing the reputed blood of Jesus Christ.[33] The circumstances by which Bruges came to have its holy relic are now uncertain but the procession of it through the city was well established by the end of the thirteenth century, a time that coincided with growth in the popularity of the cult of Corpus Christi as well as the city's own growth in wealth and influence in Flanders and beyond.[34] By 1291 'Bruges' residents [had] incorporated a vial of Christ's blood into an existing procession circumferencing their community', and perhaps had origins in 'an earlier rite venerating the Holy Cross', an indication of this being 'its May 3 celebration date'.[35]

As Boogaart makes clear, the perambulatory geography of the Holy Blood is symbolically and cosmologically significant, tracing an outline of the world in and through the city, taking in the whole city and encompassing it with the holy blood of Christ the Redeemer. The route taken by the procession is revealing in this regard for it forms a spatial link between centre and edge – inner and outer – as well as circumscribing the city by the compass of its circuit around the whole city (illus. 60). The procession began with an evening vigil on 2 May in the church of St Basil, after which, 'at the tender hour of four o'clock' the following day, the 'holy relic was taken from its tabernacle and placed on display in the Burg', the courtly and ecclesiastical centre of Bruges, where it 'remained on display until ten in the morning'.[36] There the procession gathered, the civic dignitaries, churchmen and the count's officials – the city's body politic – and with the relic safely placed 'in its carriage' and protected 'under its embroidered canopy' the procession led out from the inner Burg to the adjacent main marketplace, the Great Market (Grote Markt) with its impressive Belfort tower and civic buildings, at which point the procession was joined by the *male* citizenry at large, 'assembled by guild and parish under their banners', the social body of the city (illus. 61).[37] Here, Boogaart suggests, the tower was 'in a visual,

60 The route of the procession of the Holy Blood, Bruges (Belgium).

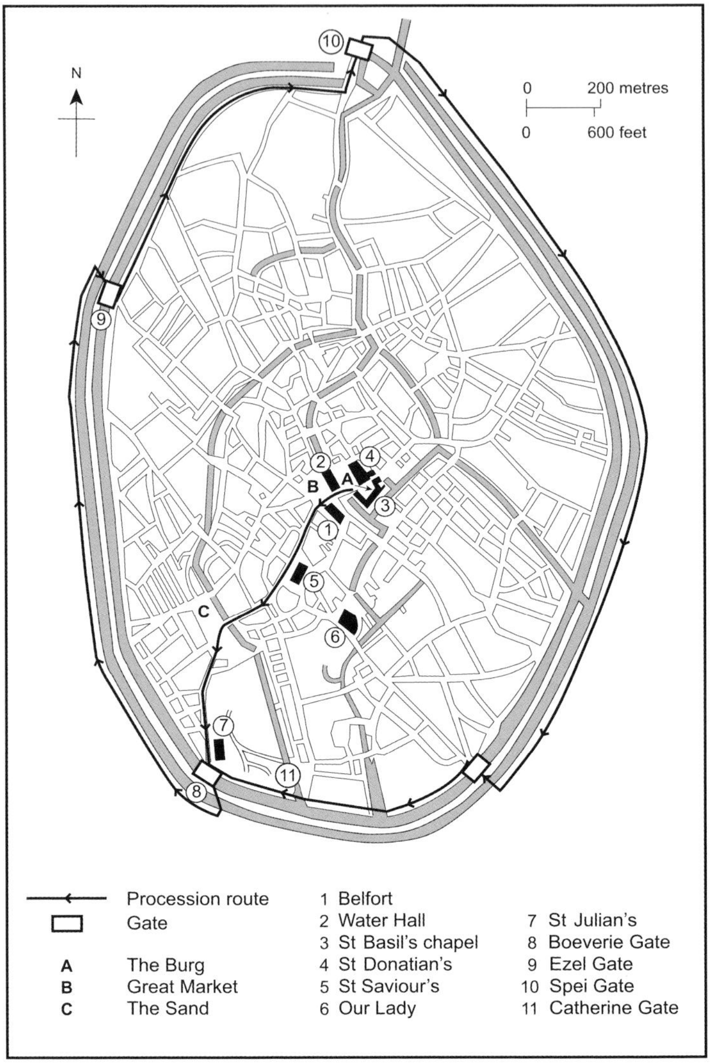

geographical and a more metaphorical sense . . . the "centre" of the community . . . an axis for denoting the symbolic correspondence of the city, community and cosmos', an *axis mundi*.[38]

The starting point of the processional route of the Holy Blood symbolized the central place occupied in Bruges by religious and political dignitaries who headed the procession, those who ordered and governed the city and tied together the places that formed its governing 'head'. The male citizenry at large, who made up the remainder of the procession, and comprising members of guild and parish bodies, joined next in the processional sequence, lower down the body and a little further out from the centre, though still part of it. From the Great Market the procession then began its course towards the city walls, moving in a southerly direction through the streets to a second, larger marketplace, the Sand, where those 'outsiders' – pilgrims, children and women – who were not part of the performing urban officiate watched as the ordered social body passed towards the Boeverie Gate at the outer edge of the city. Here the procession turned right to proceed in a clockwise direction around the whole circuit of the city walls, moving in and out through the Ezel, Spei and Catherine Gates before returning to the Boeverie Gate from which the procession returned from whence it came, back to the city's heart and returning the holy relic to St Basil's for another year.[39]

In moving through and around the whole city, the Holy Blood of Christ 'served an obvious sacramental function, fortifying the community against pestilence, natural disasters, demons lurking outside the city walls' – the spatial margins of the urban world – while 'reciting psalms at city gates enabled residents [participants] to experience the Passion by retracing the *via Dolorosa*', and so remember Christ's journey to the cross through the streets of Jerusalem.[40] 'At a certain level', notes Boogaart, 'this drew a parallel between Bruges and Heavenly Jerusalem', the procession linking the earthly city to the holy city through the Body of Christ (illus. 38 and 60).[41] The lines traced by the procession were also not without significance for they mirrored the *shape* of the wider world – the world Christ embodied – centred on the holy city (see illus. 36 and 44). In the same way that Christ's outstretched arms were seen to touch the edges of the circular-shaped Christian world, the procession defined the outer urban edge of the 'world' of Bruges, marking the city's boundary between outer 'chaos' and internal order, the latter being reflected in the carefully choreographed hierarchical sequence of the procession led through the city by Christ with his followers in order behind, the 'members' that made up the city's social body.[42]

There is therefore a social and spatial parallelism in the ordering of city and cosmos as traced out by the geography of the Bruges procession. The procession began at the city's spiritual and symbolic centre, its *axis mundi*, the place the city's ruling 'head' resided. Then with its movement from centre to edge, from inside to outside the city, and in encompassing its perimeter, the procession traced out the

61 Grote Markt and Belfort, Bruges.

moral topography of the cosmic body, with its 'purer' inner core contrasting with its outer margins, the place of the lower orders. Hence through its shared forms and hierarchical ordering, unifying the city's body yet reinforcing its divine order in the social hierarchy, the procession of the Holy Blood drew onto the city a 'map' of the cosmos.

The spatial pattern of performance evident at Bruges is also to be found in Eucharistic processions elsewhere in later medieval Europe. Those associated with Corpus Christi day itself show similar geographies and point to a shared cosmological meaning and symbolism. In 1381 at Würzburg in Germany, for example, the city's Corpus Christi procession began after mass at the sacred centre, at the cathedral where participants gathered, and from there proceeded 'through the market place to the chapel of St Gothard on the bridge over the River Main and then through the butchers' quarter to one of the city's gates, the Pleichach', from which point 'the procession would move around the city walls to another gate, the Sandertor, and pass back through it past the church of Mary Magdalen and through the church of St Augustine to the market place and cathedral.'[43] As at Bruges, then, the Corpus Christi procession at Würzburg connected those points considered as important by the urban body, places of authority and protection (illus. 62). Zika comments on the significance of this, how 'the procession symbolically linked parish and collegiate churches to the mother church, the cathedral to market place or town square, the quarters of the city each other, and to both bishop and town council', thus connecting the individual parts of the city to each other and to the urban body as a whole.[44] By 'encircling the town' there are also parallels with Rogation Day processions, a relationship 'further emphasised by the common practice of halting the procession at specific stations on the route' where 'at erected altars, antiphons were sung, followed by the opening words of the four gospels and the blessing of the host towards the four quarters of the earth', a practice with similarities to the Holy Blood procession at Bruges with the singing of psalms at the city's gates.[45]

This combination of linear and circular movement in performing Christ's body in and through the city of Würzburg is likewise comparable to Bruges, not only in tracing the same form as the macrocosm and repeating its social and spatial hierarchies but also in using rituals at specific cardinal points around its edge to orientate and align the embodied city to the wider world that lay beyond the circuit of its protective walls. Once again, therefore, there is a cosmological mirroring between the imagined form of the holy city, with its circle of walls and four-quartered orientation, and the performed processional geographies traced out by urban inhabitants.[46] The Body of Christ is central in both cities, literally and metaphorically. In imitating Jerusalem's moral topography in the living form of their own city, participants in the processions were forging a link between their own world and that which the holy city symbolized and Christ embodied.

In later medieval England too there are Eucharistic processions with significant forms, performed on Corpus Christi day along par-

62 View of Würzburg from the bridge over the River Main (Germany).

ticular routes. Some of these take linear or axial courses through the urban landscape, rather than circular patterns around perimeter walls, but similarly they link together places of symbolic importance forging a bond between Christ's body and the urban social body, and between the living embodied city and the imagined embodied cosmos. At Beverley, in the East Riding of Yorkshire, the town's Corpus Christi procession was accompanied by a series of plays performed by the townspeople, Corpus Christi plays that featured in a number of English towns and cities from the fourteenth century onwards.[47] The procession is known to have taken a line through the town, effectively dividing it into two halves, yet at the same time acting, through the sacramental symbolism of the Eucharist, as a means of socially and spatially unifying the parts that made up the 'body' of the town. It was to this end that in 1411 town officials there held 'a short conference' at which 'worshipful men of a worthier sort' came together for the purposes to discuss the Corpus Christi play performed 'to the praise and honour of God and of the Body of Christ, and for the peaceful union of the worthier and lesser commons of the town of Beverley'.[48]

These plays, and the Corpus Christi performances, were thus perceived by the body politic to help in unifying the social body of the town as a whole, as James has noted, and unifying the town with the Body of Christ.[49] The town's Corpus Christi liturgical procession was held by the 1330s, one of the earliest documented in England, but the earliest account of the accompanying pageant and plays is 1377, their designated stations of performance being recorded by 1423.[50] Six 'stations' were strung out through 'the central spine of the town', in an approximate north–south alignment, starting with one placed at the northern edge at the North Bar and then, proceeding south, at the bull ring, the north end of Saturday Market, Cross bridge, Fish Market, a location 'near the Minster, probably at the end of Highgate' and lastly 'at the beck' on the town's southern edge.[51] How the plays were performed at the six stations is a subject of controversy, not helped by the fact that no surviving texts of the plays exist, but the axial arrangement of the stations is clear.[52] The linear staging of the plays performed by the town to 'honour' the Body of Christ, along with the accompanying (and preceding) liturgical procession, helped form Beverley as one body, while the line traced out in the performance in and through the urban landscape formed an axis that, like the Body of Christ, both symbolically unified *and* divided the town socially and spatially. This 'mapping' of Corpus Christi through the particular localized geographies of performance is evident in other English towns and cities, too, forging a perceived link between Christ's body and the urban body.

As with Beverley, the historical and spatial relationship between processions and plays is not always clear, and has caused much of the disagreement among students of medieval drama over whether the plays had emerged from earlier 'liturgical' processions, or whether in fact the two had different though connected origins, one religious and one civic.[53] One particular case of this is Chester where it now seems that the 'mystery plays' were performed in an open area just outside the city walls while the procession itself ran through the streets within the walls starting from St Mary's church and then proceeding along Bridgegate Street to the High Cross, then right along Eastgate Street to finally pass through the walls and finish at the church of St John.[54] However, instead of seeing these two performances being in opposition to each other – as a division between religious and civic bodies – this arrangement could be interpreted as mutually reinforcing, for both involved the association of *corpus Christi* with the urban body as a whole,[55] a symbolic unification of their two bodies where the city and its inhabitants took part in two reciprocating displays of community and communion.

In recent years the York Corpus Christi plays also have caused much debate over how they might have been performed and what their relationship was to the liturgical procession.[56] It was also a controversial issue at the time. In 1477 the council agreed to transfer the procession to the day *after* Corpus Christi, while earlier in 1426 William Melton, a local friar, had demanded a reorganization in which procession and plays would be separated and held over two days, the plays coming first on the Wednesday before Corpus Christi day.[57] Earlier, fourteenth-century records are clear about the nature of the procession and the route it took. In 1399 an entry in the *York Memorandum Book* sets out 'the places where the play of Corpus Christi shall be played', and lists twelve stations on a route from Holy Trinity church on the city's west side to the Pavement on the east, traversing a line through the city and uniting the two halves on either side of the River Ouse.[58] Later, in 1426, the *Book* notes that 'also at that time [the citizens were] similarly making a certain solemn procession for the reverence of the body of the sacrament of Corpus Christi beginning at the great gates of the Priory of the Holy Trinity of York and this going processionally to the Cathedral Church of York, and thence to the hospital of St Leonard of York where they left the aforesaid sacrament.'[59] The stations of the play therefore followed that of the processional route for the most part, as far as the gates of the Minster church, where the two separated, the liturgical procession turning left to St Leonard's and the pageant to the right to reach the Pavement.

The significance of York's Corpus Christi routes has so far been limited to understanding how the plays were performed and whether they originated from the liturgical procession and evolved into a separate set of performances. This still remains contested.[60] However, by English medieval standards, the routes are unusually well recorded and have recently been carefully mapped out by White (illus. 63).[61] Also Mill has reconstructed the locales of each station from civic records, so there is a sound basis from which to interpret the geographies of Corpus Christi performances at York.[62] Again, the route can be seen as a linking up of important places, the churches of Holy Trinity and the Minster, the principal streets of the city, the river and bridge and the main market area of the Pavement. It also combines the two halves of the city, west and east.

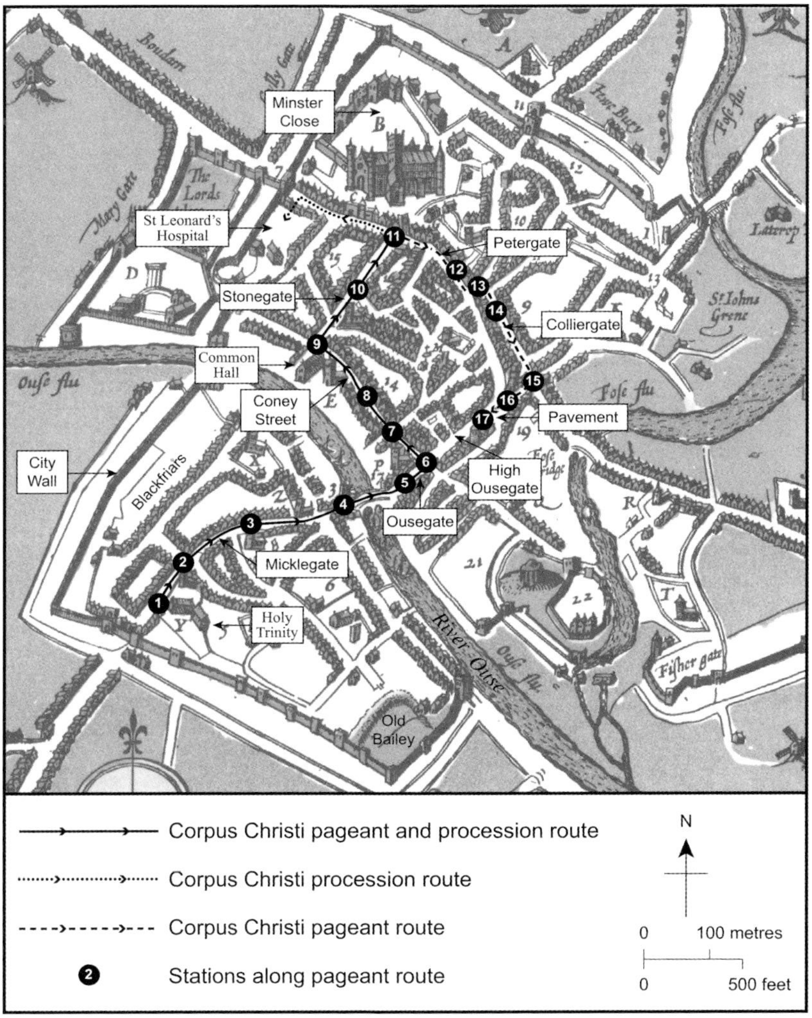

63 The route of the pageant and procession of Corpus Christi, York (England).

These are paralleled elsewhere, as we have seen, as at Beverley and Würzburg in their Eucharistic performances. The York route is also axial, or linear, and is contained within the confines of the city walls, being wholly intra-mural. Both pageants and procession began at the gates of Holy Trinity on Micklegate ('the great street'), part of a former Roman road that originally aligned with Stonegate on the east side of the river, leading to the Minster, the site of the Roman *castrum* of York.[63] Stonegate also formed part of the processional and pageant route but had to be reached from Micklegate by taking a more circuitous route through the city by crossing the Ouse by the bridge, then taking a sharp left turn along Coney Street where the Stonegate alignment is resumed by taking another sharp turn, this time to the right. Following Stonegate, the route gained the Minster gates, the point where they divided left to St Leonard's, and right along Petergate to reach the Pavement.

The form traced out by the route is a cross, the 'head' of which is the Minster. It would appear that the procession and pageant were thus initially combined as one, tracing out the Body of Christ and unifying his body and the city's body not just socially by carrying the Host through the streets, but also spatially, in forming the shape of the cross that symbolized Christ in the pattern of streets chosen for the procession and the pageants. If the aim was simply to get the pageant to the Pavement it would have been far easier to take High Ousegate rather than the curious diversion along Coney Street, Stonegate and Petergate. The reason for this must lie in the desire to keep the two routes as one – the civic pageant and the religious procession – at least as far as the Minster gate, this being the 'shaft' of the cross, Christ's body, its 'arms' then extending out in opposite directions to St Leonard's and the Pavement. As such, the processional geography traced out by the Corpus Christi routes symbolically 'mapped' his body onto the city as a whole, combining the city's social and spatial bodies and unifying them.[64] Here again it would seem that it was through enacting and performing his body in particular patterns on the ground that the city sought to imprint on itself the spatial form of the world Christ embodied.

The evidence for York's Corpus Christi processional route is not only significant for what it reveals about the ways the urban social body performed a projected embodiment of itself, in the way that James and others have explored, but also in the performed geographies that reflect a cosmology in which Christ's body was also central. Important and celebrated Corpus Christi plays and pageants were enacted by and for the corporeal city elsewhere in England at this time, to unify yet divide its body for 'peaceful union' of the whole 'in honour of the Body of Christ'. Unfortunately, their routes are not so easily traced from written accounts, as at Coventry, where although various stations for performing pageants are mentioned in civic records of the fourteenth and fifteenth centuries, there is no way of linking them to form a single processional route even though such a route did exist.[65]

In continental Europe, in places such as Bruges and Würzburg, the evidence is perhaps clearer than it is for England, not merely in

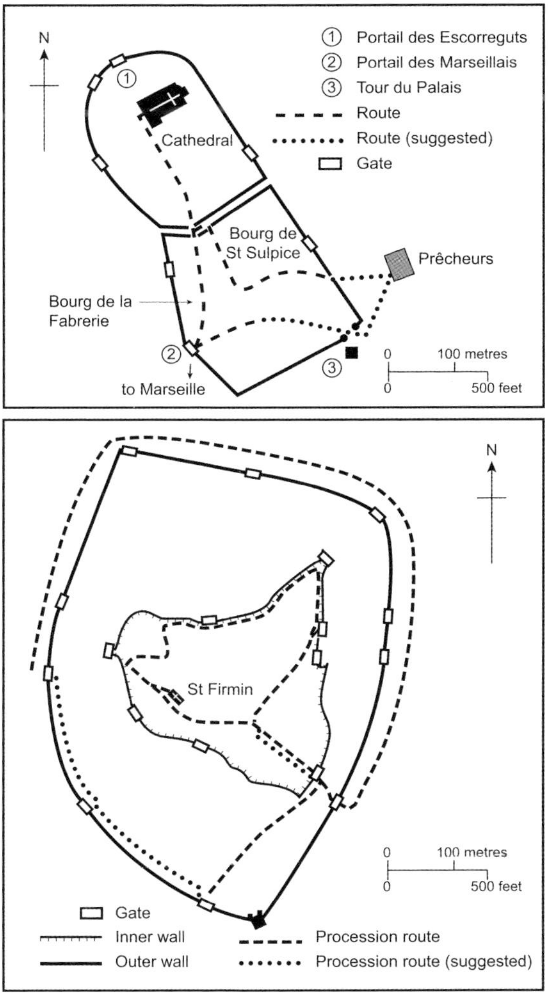

64 Processional routes in southern France: Le Fête-Dieu, Aix-en-Provence, fourteenth century (top), and Montpellier processional route of 1407 to deliver the city from pestilence, circumscribing the inner and outer walls of the city (below).

the case of Eucharistic processions but other ritual urban performances that likewise traced out geographies in and through the urban landscape, bringing together as one the social and spatial bodies of the city and imprinting upon them a cosmological symbolism. For example, Coulet has mapped out similar performed geographies in medieval Marseilles, Aix and Montpellier, and these again show routes that order and organize urban spaces and bodies, sometimes through venerating local saints or the apostles, sometimes through Corpus Christi processions (illus. 64).[66] Their spatial patterns and forms are familiar ones, linking places within and outside the city, encircling its outer walls, weaving in and out of gates, traversing principal roads and marketplaces in axes that cross and orientate the earthly urban world in ways that imitate the patterns and forms of the cosmic body at large (illus. 45 and 46). These geographies of performance and the traces of the world they formed provided a living link between city and cosmos, imprinting an 'image of the world' upon the city whose participants commemorated through their bodies the body of their Saviour.[67] In this regard, their performances were doing more than tracing the outlines of the world he embodied, for they were also performing his body as a re-enactment of cosmic history, or cosmogony, since Christ's body was understood as a 'map' of the world's creation, salvation and judgement (illus. 44).

Embodiments of cosmogony

The rituals in the Holy Blood procession at Bruges 'bridged cosmic and civic history at different levels, implicitly associating community and cosmos, world salvation and the restoration of justice during historical time', embodying a conception of Christian cosmogony.[68] Here this theme is explored in part by looking at Boogaart's thoughts on the Bruges procession and its cosmogenic symbolism, but also more broadly by looking again at Eucharistic performances and examining their connections with sacral history, how the social body of the city, through performing Christ's body, was also performing the world God made, for in the Body of Christ was the history of the world, its past, present and future. Two aspects of this are examined here, first the performance of cosmic *time* within particular urban spaces, and second the performance of cosmic *space* in the city's ritual calendar. In both there is a convergence of cosmic and urban space-time. Both appear in the plays and pageants that occurred in or around Corpus Christi day in towns and cities across Europe in the later Middle Ages, and both forged a symbolic connection between city and cosmos.

In her analysis of 'Corpus Christi drama', Rubin notes how 'the staging of sequences of biblical dramatic plays . . . combined the processional, the didactic, the civic, the competitive, the creative and the fantastic' in an 'array of configurations which only a cosmic theme could bring about'.[69] The plays and pageants that urban bodies performed 'in honour of the Body of Christ' took many different forms. Sometimes they were *tableaux vivants* within or alongside the liturgical procession, at other times static 'plays' were performed on specially erected stages, scaffolds or 'pageants' in particular locations, or occasionally there were 'processional plays' by which the performances were held on mobile pageant wagons and moved through the urban landscape.[70]

The orthodoxy that Corpus Christi plays of the fourteenth and fifteenth centuries evolved from earlier liturgical processions is now suspect, as is the view that 'processional plays', performed on moving wagons, were the norm, at least in the context of late medieval England.[71] Certainly they were popular throughout Europe, especially in England and Germany, where their prevalence is particularly evident.[72] The English plays were performed in the vernacular, by local townspeople, and were usually arranged, funded and acted by craft guilds who took responsibility for their own individual play which formed part of an overall larger, longer sequence, or 'cycle', overseen by civic officials. This cycle was therefore the 'creation of its community, an expression of civic pride and piety', performed by the urban body for the urban body, and like the Corpus Christi procession was an act in which the members that made up the embodied city as a whole unified themselves with each other through communion with Christ's body.[73] The mutual purpose of the York play was made clear in an ordinance issued in 1417: 'all the pageants of the play called Corpus Christi play should be supported and produced by the artisans of the said city in their order, for the due honour and reverence of our Lord Jesus Christ and the benefit of the said citizens.'[74] As Nijsten has observed, such 'feasts, processions and dramatic spectacles provided the town with a language and grammar, as it were, to express this union of differences', and with their 'strong emphasis on the omnipotence of God, who had created the world and had established its unchanging order', this 'was an order that served the rulers well' in governing those bodies that made up the city.[75]

The Corpus Christi plays performed in late-medieval English towns and cities 'portray the history of the world from its creation until its end.' [76] It is a 'cycle that tells the Christian story of the world from its beginning to the end of time . . . opening before God's creation of the world and closing after the final judgement of mankind', in short, a cosmogony.[77] It is the same sacramental history visualized by *mappaemundi*, likewise derived from Scripture, from both Old and New Testaments (see chapter One) (illus. 36 and 44). Thus, in the case of York, for example, the first of the plays is on the Creation, and last is Judgement Day, and in between Christ's life is narrated, including his entry into Jerusalem, his crucifixion and his resurrection, the history of the world being embodied by Christ *and* foretold by scripture.[78] This same pattern was also followed in Corpus Christi plays performed at Coventry and Chester, and indeed elsewhere.[79] These embodied performances of Christian cosmogony took place in and through the urban landscape at prescribed places that varied in number, both between towns and also over time within the same town. Thus at York the number of stations where performances took place ranged from twelve in *c.* 1400 to sixteen in *c.* 1550.[80] Even so, certain key places were consistently used as 'stations' across this period, including the Holy Trinity gates (the start point), the 'Common Hall', the Minster gates, and the Pavement (always the final station) (illus. 63).[81] Indeed, in 1394 'the mayor, bailiffs and commonalty' agreed 'that all the pageants should play in the places appointed of old time [*antiquitus assignatis*] and not elsewhere.'[82] What was actually performed at each of these stations however is again a matter of debate; some perhaps were shortened versions of the plays themselves, others lengthier pieces, with the pageant wagons moved from station to station around the city.[83]

It is because the Corpus Christi play 'cycles' would appear to have been too long to have all been performed processionally (48 known plays at York), and logistically too complex (at up to 12 or more 'stations' in a single day), that doubt has been cast on whether cycle plays such as those at York were performed 'in the "true-processional" manner.' [84] Instead the suggestion is that in York at least shorter versions of the plays were enacted as part of the procession, stopping at the stations *en route*, with a later subsequent longer cycle performance of the plays (or selected plays) perhaps at the Pavement or in a chamber in the 'Common Hall', the civic chambers.[85] Either way, the plays' content, with their emphasis on cosmic time, was connected with particular spaces within the city (illus. 65). The Corpus Christi play was thus a dramatic localized display of world history, 'a dramatic species in cosmic form'.[86] Again, it was Christ who symbolized this history, the past (creation), present (salvation) and future (judgement) of the world, and by performing these plays on Corpus Christi day the urban body was connecting itself, the spatial and social embodiment of the city, to the wider world.

Whether or not Corpus Christi plays evolved from the liturgical procession of *corpus Christi*, the two are certainly symbolically connected through located and embodied performances of his body. Although the dramaturgical content of the cycle arose not from the liturgy of the feast of Corpus Christi itself, 'the fact is . . . separate,

65 Performing the York 'mystery play', July 2002.

preexisting plays associated with the Nativity and with Easter were grouped into a single cycle of cosmic scope.'[87] This connection between the city (in urban embodiment and performance) and cosmos (in scriptural history and Christ's body) was forged by the procession of the Host *and* the plays that took place on Corpus Christi day. Through them 'the idea of space was joined with that of time to make a total unity of time, space, and experience' for those living within the city.[88] There was, then, a convergence of cosmic time and urban space in performing Corpus Christi plays. Indeed, as Karhl puts it, 'conceptions of time developed in the early Christian period' were 'later embodied in the English mystery cycles', and 'nowhere can medieval notions of cycles of history be seen so well as in the civic plays which portray the history of the world from its creation until its end.'[89]

The physical movement of pageants in processional performances of plays symbolized the passage of time, as well as the plays' contents, which narrated the story of the world from Creation to 'Doomsday'. Cosmic time was thus being written into the urban landscape by the urban body performing the world. At York, 'the actual space of the Pavement mattered in the staging of the Last Judgement', for it was not only the final play that marked the end of all time but also the final station in the whole cycle. The end was therefore marked both spatially and temporally in this final sequence, performed before the city at its civic heart, in a play that reminded those who made up the urban body, the plays' viewers and participants, of what was to come: the division in society between those who will be saved and those eternally damned.[90] The places of performance in and through the city that were repeatedly used in presenting these plays would over time become ever associated with their cosmogenic narrative and moral meaning. This association between cosmic time and urban space would be especially clear to those whose guilds were performing the individual plays, those whose collective bodies were each forming a part of the entire Christian story, whose particular crafts were to 'relate to the subject of their scenes', as at York, Chester and Newcastle.[91] Just as the parts of the cycle connected together to make up the cosmological sequence, so their individual and combined 'members' were connected together through the plays to perform the history of the world. It is for this reason perhaps that the ritual mimetic enactments of Christian cosmogony that formed these plays were placed alongside, and in some cases made a part of, a commemoration of *corpus Christi*, which through the annual procession of the Host likewise took in the whole body of the city: in both it was the Eucharist that performed this connection between city and cosmos.

The central role of the Eucharist in linking cosmic time and urban space is of course not confined to the Corpus Christi plays performed in later medieval English towns and cities. It also performed this purpose elsewhere for such plays were commonplace in other European urban centres, in the Low Countries, in France, Germany, Spain and Italy.[92] Neither is it confined to the feast of Corpus Christi, as Boogaart makes clear in his discussion of the procession of the Holy Blood at Bruges: 'in processing the Host through urban space and reciting psalms at different urban landmarks, residents identified the parallelism between cosmic history and their habitat' within the city.[93] It is the world history embodied by Christ – creation, salvation, judgement – that makes the phial of his Holy Blood a point of contact between the 'cosmic and civic history' in Bruges, but also the movement of the relic in and through the city (like the processional cycle of Corpus Christi plays) that connected cosmic time and urban space, for the procession itself 'appears to have originated as a rogational right' in which 'perambulation of communal boundaries' took place and 'recitation of benedictions at cardinal points reoriented communal life with its invisible sacred pivots, reharmonizing the community with biocosmic rhythms.'[94] The Holy Blood procession thus related the city to the history of the world in two connected ways, through its embodiment of cosmic time by moving the Host through the city and around it, and through the annual enactment of cyclical time at a key point in the city's ritual year, with the performance of an annual rite encircling the city by the whole community. In both, cosmic time is spatialized by the performing urban body (illus. 60).

The coincidence between the timing of the chosen date for the feast of Corpus Christi and pre-existing summer-time rites and rituals has been noted elsewhere.[95] It is the physical movement of Christ's body in particular linear and circular (clockwise) patterns in urban space that signifies the passage of cosmic time, whether at the universal (linear) scale, covering Creation to 'Doomsday', or at the local (cyclical) scale, within the city's Christian calendar – in both 'the ritual operated as a "meta-narrative" communicating the intrinsic parallelism between the civic community and the cosmic order.'[96] Other festivals in the civic year were also cyclical and 'sought to invoke divine blessing on the material world', as Schriber notes, and were also a means by which 'cosmic order and daily life' were connected in and through the city.[97] As well as these annual cycles, the procession at Bruges also connected the city's own (linear) history with the cosmic history its performance embodied, for the phial of Holy Blood was understood by the urban community to protect them and their rights in the face of adversity, so much so that that when Bruges surrendered to Philip the Fair of France in 1297 it did so 'on the singular condition that he [did] not remove the precious relic from their city.'[98] To lose this link with Christ's body was to lose not just their city but their past, present and future.

The focus so far has been on how Corpus Christi plays in particular and Eucharistic processions in general provided a means by which urban bodies performed the world, embodying cosmic time and spatializing Christian cosmogony in and through the urban landscape. These performances were also embedding cosmic space in the city, performing another living link therefore between city and cosmos. As well as having a temporal place in the civic year, plays and processions that annually celebrated Corpus Christi were creating in the earthly city simulacra of the heavenly city, placing in the everyday material urban world symbols of the wider world that Christ embodied. Again, these symbols were associated with cosmic events, as foretold by Scripture and in Christ's life, and provided townspeople with a transcendental experience in which both cosmic space and time came together in the space and time of their own urban existences. The staging of plays makes this clear.

Where they exist, descriptions of the performances of plays reveal how at certain times of the ritual year the urban landscape was transformed into celestial places with mystical and divine meaning. The cycle plays performed at Chester, for example, were held just outside the city's walls in an area called the Rood-Eye, chosen according to Powlick because of its suitability as a place of performance for the plays' staging, in particular the proximity to the city wall and gates needed for example in the seventh play, 'The Adoration of the Shepherds'.[99] The Chester cycle was itself a complex development of Old Testament and Nativity scenes that evolved between the early fifteenth century and the sixteenth century.[100] In its later guise as a stationary sequence of plays the dramatic story of the 'mystery cycle' was placed in a setting which gave realism to the scenes being performed, such as the thirteen out of twenty-four plays that required a hill, a 'high place', on which heaven was situated and Christ crucified, for example.[101]

Such local placing of scenes at particular landmarks gave locales allegorical meaning, performing a link between them and their cosmic archetype, whether as a representation of heaven, the celestial space, or events such as the crucifixion with their divine meaning in Christian cosmogony. Such placing mirrored heavenly and earthly worlds, a reflection and reinforcement of God's 'divine power' and omnipotence which the Chester cycle is understood to have portrayed.[102]

Elsewhere too, performances connected the world above with the world below through the placing of celestial symbols within the material urban landscape. Another enactment of Christ's life was played out at Whitsuntide at Frankfurt, where a central square in the city, the Samstegberg in 'the eastern part of the Römer', formed a backdrop.[103] 'Exemplary topographical detective work' has revealed the stage-plan's layout and content.[104] It was laid out regularly in a rectangular form, and had a particular 'moral topography'. The central sites narrated the story of Christ's birth, crucifixion and resurrection, with the positions of Paradise and Jerusalem occupying particularly significant places towards the east end of the square, at its 'head', so to speak, with the Temple at the very centre of the arrangement. In this ordered form, the places Christ knew and the events his life signified were materialized in the heart of Frankfurt. The performance also drew parallels between the city and Jerusalem by a procession on the second day in which Christ 'was dragged as a prisoner "*per multas vicus civitatis*"' and a further procession the following day in which 'Christ was once again led through the streets of Frankfurt' before the play resumed, with a final procession on the last day 'which led its participants through the Sachsenhausengate in order to erect three crosses outside the city wall.'[105]

At Lucerne, too, a rectangular layout of open stages was used for a Passion Play at the end of the sixteenth century held in the Weinmarkt, for which contemporary plans survive. It shows similarities in form to Frankfurt's plan, with the 'head' of the stage orientated towards the east. Here Heaven was located, 'high up, at the top of the square, raised between two oriels of the *Haus sur Sonne*', while the Temple was placed on the south side of the square.[106] A similar concern for 'positional symbolism' of local features is to be found at York. In the plays performed on the Pavement at the final station of the Corpus Christi cycle, 'the representation of the Last Judgement', hell was 'to stage left, toward the stinking Shambles, and paradise to the right, near the church of All Saints, the burial place of the mayors'.[107] Indeed, it was at York in 1417 that occupants of properties along the processional route paid the civic authority for the privilege of having the plays staged 'before the doors and houses', and gained from it honour and standing.[108]

In all of these examples an analogy was created between the order of the heavenly world and the order of the places in which the plays are performed, while at the same time the performances within these places, in celebrating Christ's life and narrating scriptural history, were re-enactments of Christian cosmogony. Together they performed the world through embodying cosmogony, as told by divine scripture and as personified in Christ. Local urban features were thus associated with cosmic counterparts. That these mimetic acts that sought to conflate the civic year with times and places of cosmological importance were understood to have such meaning by their participants is suggested by the words of Margery Kempe while she stood watching a procession on Palm Sunday in which it seemed to her 'gostly sygth' as though she had been at Jerusalem 'and seen owr Lord in hys manhood receyuyd of the pepil as he was whil he went her in erth'.[109] Such a spiritual bond between earthly and heavenly bodies were what performances of Christ's body were seeking to materialize in and through the urban landscape in plays, pageants and processions of *corpus Christi*, mirroring in the earthly city the form and history of the world above, and so creating a living link between city and cosmos. Through Christ's body, and its perceived embodiment of cosmic form and history, the urban body as a whole, though riven with conflicts and tensions, presented a show of unity as well as difference, for a body both unified and differentiated: thus 'the ruling elite's model *of* reality was also a model *for* reality in that they used it to shape the town to their own idea of what a town and its citizens should be.'[110] In so doing they shaped their own urban body into an image of the cosmic body.

EPILOGUE

CITIES OF GOD?

So what shall we say of the city? Today, at a time when the world appears to be riven with religious and cultural difference and as science and religion are seen to be opposing rather than unifying forces, looking at how the city was understood by Christians in the Latin West may seem somewhat esoteric. But instead it reveals historical precedent for what can be achieved through religious, scientific and artistic exchange, even in an age viewed conventionally (and in my view unfairly) as equally dogmatic, warring and intolerant.

Europe in the Middle Ages was in fact a fertile place for human endeavour, of making progress in the spiritual and material realms of the living. It is through this industry that Europe was made. For those who conceived, built and inhabited its towns and cities – those who described them and depicted them, who created them and lived in them – the city was a special place, for it was a model of the world that was to come, a city of God. They read this from Scripture, saw it illuminated in their religious art and iconography, heard it in sermons and in speeches and performed it in their rites and devotion.[1] This imagined 'city', symbolizing the celestial, the heavenly and the wider world, held a powerful grip on medieval minds. It took shape through theological and scientific engagements with God and the created universe, in trying to make sense of this world and the next. In Christian thinking this city-cosmos idea(l) pointed towards God. No wonder Augustine used it to argue his case 'concerning the city of God against the pagans', for the city was itself a motif of salvation connecting God with 'man', the spiritual with the material, the heavenly with the earthly. In short, the earthly city and the heavenly 'city' were cities *of* God.

The cosmological symbolism of the medieval city and the idea that through its form the city was an imitation of the wider world and its past, present and future is peculiar neither to Europe nor to Latin Christianity. Far from it in fact, for the symbolism of the city as it was understood by medieval Europeans was shared among those whose ideas had played a role in shaping Christian culture in the Middle Ages, including Judaic and Islamic art, theology and science, and the 'Classical' natural and political philosophy of the ancient Mediterranean world, in particular Platonism and Aristotelianism.[2] These influences on how the city was understood in the Latin West deserve closer scrutiny, especially with other monotheistic cultures of the time, for parallels exist between those who perceived the city as a means of mediating between worlds, a symbol by which God's presence could be felt.

Christianity is fundamentally an urban religion, the city having a prominent place not just in its history, doctrine and beliefs, but in its organization, practice and worship.[3] But it would be worth also looking at how its modes of urban representation, those images that depict the city geometrical in shape, a symbol of the cosmos, are to be found elsewhere, for example in Hinduism in mandalic forms where geometry is likewise used to draw out symbolic parallels between city and cosmos.[4] So too with the material forms of urban landscapes in medieval India and imperial China, for example, where a sacred geometry was used not only to define the layouts of towns and cities on the ground, but to orchestrate their formation, for deciding on urban location, siting and shape.[5] There is good evidence of shared beliefs and practices between ancient European, particularly Roman and Etruscan, rites of urban foundation and those of the Indian sub-continent.[6] Through copying *Corpus Agrimensorum* texts into gromatic treatises, medieval scribes were linking earlier Indo-European urban traditions with those of the Latin West. The micro-macrocosmic thinking that Alan of Lille, Thomas Aquinas and other revered Christian theologians of the Middle Ages used in composing their works was also important in forming, mediating and articulating perceived relationships between earthly and heavenly worlds in many different cultural contexts, including the pre-Columbian Americas, as well as Asia.[7] It features, for example, in the civic rituals of Hindu Nepal, in the annual rites that are celebrated in the Newar city of Bhaktapur where the symbolic meeting of the cosmic body of the wider universe, and the microcosm, the human body, is performed in 'calendrical events' in the inhabitants' 'festival year', in rituals and local enactments in certain times and spaces that bear very close resemblance to the Eucharistic re-enactments and processions in towns and cities of medieval Christian Europe.[8]

Such parallels between city and cosmos in various religious contexts are evidence for interdenominational and cross-cultural exchange in the past, and through their continuing study may help point to the shared principles and practices of faith that unite rather than divide the world today. The city of God, as it was conceived by the Latin West, still has a strong resonance even a millennium later. It has the potential to provide a common ground, a meeting point for Abrahamic and other religions, for it is part of a shared cultural inheritance that has a long and influential place in forming human experience, both material and spiritual, which only now is beginning to get pieced together through comparative and interdisciplinary study. To this end, what is offered here in this book may provide a small contribution towards this ongoing debate, and help foster further understanding between those searching for their own particular city of God.

REFERENCES

INTRODUCTION

1 H. Bettenson, ed. and trans., *Augustine, The City of God* (London, 1984), p. 859.

2 A. J. Gurevich, *Categories of Medieval Culture*, trans. G. L. Campbell (London, 1985), p. 59.

3 The Latin 'homo' most simply translated as 'man'. I am conscious of the gender implications of this. On the relationship between ancient and medieval micro- and macrocosm more generally see G. P. Conger, *Theories of Macrocosms and Microcosms in the History of Philosophy* (New York, 1922), pp. 7–11, 20–23, 29–36; R. Allers, 'Microcosmus: From Anaximandros to Paracelsus', *Traditio*, II (1944), pp. 319–407.

4 D. Lee, ed. and trans., *Plato, Timaeus and Critias* (London, 1977), p. 29. The *Timaeus* is usually taken to be 'one of the last works which Plato wrote', following the *Republic* and the *Laws*, pp. 22–3.

5 On translations and circulation of Greek philosophical and scientific works in the later Latin West see for example C. H. Haskins, *Studies in the History of Medieval Science*, second edition (Cambridge, MA, 1927); J. T. Muckle, 'Greek Works Translated Directly into Latin before 1350', *Mediaeval Studies*, IV (1942), pp. 33–42, 102–14; M. Gibson, 'The Study of the *Timaeus* in the Eleventh and Twelfth Centuries', *Pensiamento*, XXV (1969), pp. 183–94; D. C. Lindberg, 'The Transmission of Greek and Arabic Learning to the West', in *Science in the Middle Ages*, ed. D. C. Lindberg (Chicago, IL, 1978), pp. 52–90; R. Klibansky, *The Continuity of the Platonic Tradition*, new edn (New York, 1982), pp. 13–54. On Aristotle's cosmos see E. Grant, *Physical Science in the Middle Ages* (Cambridge, 1971), pp. 60–82. For Aquinas on *Physics* see R. J. Blackwell, R. J. Spath and W. E. Thirkel, trans., *Commentary on Aristotle's 'Physics': St Thomas Aquinas* (Notre Dame, IN, 1999). On Aquinas' *Politics* see P. E. Sigmund, ed. and trans., *St Thomas Aquinas on Politics and Ethics* (New York, 1988); J. Dunbabin, 'The Reception and Interpretation of Aristotle's *Politics*', in *The Cambridge History of Later Medieval Philosophy: From the Rediscovery of Aristotle to the Disintegration of Scholasticism 1100–1600*, ed. N. Kretzman, A. Kenny and J. Pinborg (Cambridge, 1982), pp. 723–37; C. Martin, 'Some Medieval Commentaries on Aristotle's *Politics*', *History*, XXXVI (1951), pp. 29–44. On Aquinas generally see B. Davies, 'Thomas Aquinas', in *Routledge History of Philosophy, III: Medieval Philosophy* , ed. J. Marenbon (London, 1988), pp. 241–68.

6 W. H. Stahl, *Roman Science: Origins, Development and Influence to the Later Middle Ages* (Madison, 1962), p. 142. On the later influences of Calcidius' commentary and translation see particularly W. Wetherbee, 'Philosophy, Cosmology and the Twelfth-Century Renaissance', in *A History of Twelfth-Century Western Philosophy*, ed. P. Dronke (Cambridge, 1988), pp. 21–53; T. Gregory, 'The Platonic Inheritance', in *History of*

Twelfth-Century Western Philosophy, ed. Dronke, pp. 54–80; also Gibson, 'Study of the *Timaeus*' and R. Klibansky, *The Continuity of the Platonic Tradition*, new edn (New York, 1982). See R. W. Southern, *Platonism, Scholastic Method and the School of Chartres* (Reading, 1978), especially pp. 5–15, and his graph showing surviving medieval manuscripts of the *Timaeus* over fifty-year periods from 900–1500 (p. 14).

7 Klibansky, *Continuity of the Platonic Tradition*, p. 28. 'At least seventy-seven manuscripts of the work survive from those two centuries [the eleventh and twelfth], evidence of a remarkable proliferation and diffusion of copies throughout Europe', writes R. D. Crouse, '*Hic sensilus mundus*: Calcidius and Eriugena in Honorius Augustudunensis', in *From Athens to Chartres: Neoplatonism and Medieval Thought. Studies in Honour of Edouard Jeauneau*, ed. H. J. Westra (Leiden, 1992), pp. 283–8, at p. 283. There was a copy of Calcidius' *Timaeus* at Saint Gall and Lyon in the ninth century, but at that time 'the *Timaeus* seems to have been a venerated curiosity rather than a work that men used and understood', the real development in its dissemination and use coming later, in the eleventh and twelfth centuries, according to Gibson, 'Study of the *Timaeus*', pp. 183–4, 190, at p. 184 (and for her list of medieval Timean manuscripts see pp. 185–8).

8 For example see the debate between Peter Dronke and Paul Dutton on the possible authorship by Bernard of Chartres of a gloss on the *Timaeus*, P. Dronke, 'Introduction', in *Twelfth-Century Western Philosophy*, ed. Dronke, pp. 14–18, and P. E. Dutton, ed., *The Glosae super Platonem of Bernard of Chartres* (Toronto, 1991), pp. 8–21. On the vexed issue of the 'school' of Chartres see Southern, *Platonism, Scholastic Method and the School of Chartres*; R. W. Southern, *Scholastic Humanism and the Unification of Europe* (Oxford, 1995), pp. 59–101.

9 P. E. Dutton, '*Illustre civitatis et populi exemplum*: Plato's *Timaeus* and the Transmission of Calcidius to the End of the Twelfth Century of a Tripartite Scheme of Society', *Mediaeval Studies*, XLV (1983), pp. 79–119, at pp. 83–4. Referring to the passage in Lee, ed. and trans., *Plato, Timaeus*, p. 61.

10 Quoted from Dutton, '*Illustre civitatis et populi exemplum*', p. 84, citing J. H. Waszink, ed., *Timaeus a Calcidio translatus commentarioque instructus*, 2nd edn (London, 1975), p. 246.

11 Quoted from Dutton, '*Illustre civitatis et populi exemplum*', pp. 84–5, citing Waszink, ed., *Timaeus*, p. 247. Dutton explains that Calcidius is linking the conceptual schema of the city-state from Plato's *Republic* with that of the body and cosmos in the *Timaeus*.

12 Dutton, '*Illustre civitatis et populi exemplum*', pp. 86–117.

13 J. J. Sheridan, trans. *Alan of Lille: The Plaint of Nature* (Toronto, 1980), p. 120. See also Dutton, '*Illustre civitatis et populi exemplum*', p. 114.

14 Sheridan, trans. *Alan of Lille*, pp. 121–2. See also Dutton, '*Illustre civitatis et populi exemplum*', pp. 114–17.

15 Dutton, '*Illustre civitatis et populi exemplum*', p. 94, with the passage taken from p. 93, citing É. Jeauneau, ed., *Guillaume de Conches, 'Glosae super Platonem'* (Paris, 1975), p. 75. On William's *Timaeus* see also Southern, *Platonism, Scholastic Method and the School of Chartres*, pp. 16–25.

16 Cited in Dutton, '*Illustre civitatis et populi exemplum*', p. 92. For Macrobius see Stahl, *Roman Science*, pp. 151–69; W. H. Stahl, trans. *Macrobius' Commentary on the Dream of Scipio* (New York, 1952).

17 Typically the history of European medieval urbanism is presented as

one of ecomonic and political progress, rather than a philosophical and cultural phenomenon; see H. Pirenne, *Medieval Cities: Their Origins and the Revival of Trade*, trans. F. D. Halsey (Princeton, NJ, 1925); F. Rörig, *The Medieval Town*, trans. D. Bryant (London, 1967); E. Ennen, *The Medieval Town*, trans. N. Fryde (Amsterdam, 1979); R. H. Hilton, *English and French Towns in Feudal Society: A Comparative Study* (Cambridge, 1992).

ONE: URBAN MAPPINGS

1 On Jerusalem being the *axis mundi* see P. Wheatley, *The Pivot of the Four Quarters: a Preliminary Enquiry into the Origins and Character of the Ancient Chinese City* (Edinburgh, 1971), p. 428. He notes that in the twelfth century 'the Icelandic pilgrim Nicholas of Thverva . . . reported that at Jerusalem (which was built on the rock that constituted the navel of the earth) "on the day of the summer solstice the light of the sun falls perpendicularly from Heaven"', citing L. I. Ringbom, *Graltempel und Paradies* (Stockholm, 1951), p. 255. Similarly, 'Adomnán, abbot of Iona, in his *De locis sanctis*, speaks of: "a very high column which stands in the centre of the city . . . It is remarkable how this column . . . fails to cast a shadow at midday during the summer solstice, when the sun reaches the center of the heavens . . . And so this column . . . proves to be at the center of the world . . . and its navel"', D. Woodward, 'Medieval *mappaemundi*', in *The History of Cartography*, vol. I: *Cartography in Prehistoric, Ancient and Medieval Europe and the Mediterranean*, ed. J. B. Harley and D. Woodward (Chicago, IL, 1987), pp. 286–370, at p. 340, citing J. B. Friedman, *Monstrous Races in Medieval Art and Thought* (Cambridge, MA, 1981), p. 219, n. 23. Adomnán's descriptions of holy places had been supplied by the French bishop Arculf in about 679–80, and included 'four sketches of the buildings of Jerusalem', done by Arculf himself after visiting the city; see C. Frugoni, *A Distant City: Images of Urban Experience in the Medieval World*, trans. W. McCuiag (Princeton, NJ, 1991), p. 19 and also J.J.G. Alexander, '"Jerusalem the Golden": Image and Myth in the Middle Ages in Western Europe', in *The Real and Ideal Jerusalem in Jewish, Christian and Islamic Art*, ed. B. Kühnel (Jerusalem, 1998), pp. 254–64, at p. 255, n. 7. Cf. Ezekiel 5:5.

2 P.D.A. Harvey, 'Local and Regional Cartography in Medieval Europe', in *History of Cartography*, I, ed. Harley and Woodward, pp. 464–501: 'most medieval maps of Jerusalem – and many dozens survive – give it a circular form' (p. 474).

3 The Psalter map is reproduced in colour in B. Kühnel, 'The Use and Abuse of Jerusalem', in *Real and Ideal Jerusalem*, ed. Kühnel, pp. xix–xxviii, fig. 5. Woodward, 'Medieval *mappaemundi*', p. 325, refers to 'the world map of Andreas Walsperger (1448)' who 'explains his system of distinguishing between Christian and Islamic cities: "The earth is indeed white, the seas of a green colour, the rivers blue, the mountains variegated, likewise the red spots are cities of the Christians, the black ones in truth are the cities of the infidels on land and sea"'. He also notes (p. 326) that 'the Arabic world maps are generally more abstract [than Christian ones] in their use of signs, using circles for cities'. See also P. Laveden, *Représentation des villes dans l'art du Moyen Age* (Paris, 1954), p. 12: 'dans la *Mappa mundi* d'Hereford, au coeur d'un monde

circulaire, Jérusalem est elle-même une ville circulaire, dont l'enceinte crénelée est régulièrement flanquée de quatre tours et percée de quatre portes'.

4 R. Ousterhout, 'Flexible Geography and Transportable Topography', in *Real and Ideal Jerusalem*, ed. Kühnel, pp. 393–404.

5 Revelation 21:2 and also 21:10.

6 The otherwise excellent medieval volume of the Chicago *History of Cartography* largely overlooks the medieval imagery of the Heavenly Jerusalem, which is unfortunate. 'An art-historical study on the Jerusalem maps and views is still a *desideratum*', observes Kühnel, 'Use and Abuse of Jerusalem', p. xxii, n. 12. On the origins of the imagery see I. Katz, 'Les représentations de villes dans l'art Chrétien avant l'an mil', *Information de l'histoire de l'art*, III (1964), pp. 130–32, at p. 131; on Carolingian examples see M-T. Gousset, 'La représentation de la Jérusalem céleste à l'époque Carolingienne', *Cahiers Archéologiques*, XXIII (1974), pp. 47–60, and Frugoni, *Distant City*, pp. 3–29; on later examples see P. H. Brieger, *The Trinity College Apocalypse: An Introduction and Description* (London, 1967). An overview is M. Levy-Rubin and R. Rubin, 'The Image of the Holy City in Maps and Mapping', in *City of the Great King: Jerusalem from David to the Present*, ed. N. Rosovsky (Cambridge, MA, 1996), pp. 352–79; and B. Kühnel, *From the Earthly to the Heavenly Jerusalem: Representations of the Holy City in Christian Art of the First Millennium* (Rome, 1987).

7 Gousset, 'Représentation de la Jérusalem', pp. 48–56; Frugoni, *Distant City*, pp. 20–23. On the 'tree of life' see G. B. Ladner, 'Medieval and Modern Understanding of Symbolism: A Comparison', *Speculum*, LIV (1979), pp. 223–56.

8 See Brieger, *Trinity College Apocalypse*, pp. 8–11. He refers to 'the northern group of Apocalypses' (p. 11) in which the Heavenly Jerusalem is shown in elevation, and comments also on the 'Beatus tradition' (in which he places the Trinity example) where the city is shown 'in projection', where 'its four walls are laid flat on the ground against the sides of the centre square upon which the gates rest'. The square-shaped, planimetric image of the Heavenly Jerusalem of the Morgan Beatus manuscript of AD 922–6 contrasts with its projection of Babylon, shown in 'frontal elevation' to signify the 'moral difference' between the two cities according to A.W.G. Posèq, 'The "New Jerusalem", the Star of Zion and the Mandala', in *Real and Ideal Jerusalem*, ed. Kühnel, pp. 325–39, at 328, figs 1 and 3.

9 Revelation 21:13–16.

10 Gousset, 'Représentation de la Jérusalem', p. 54, figs 1–2, 4, 6–10, 14; Frugoni, *Distant City*, p. 13, plate 16; Brieger, *Trinity College Apocalypse*, pp. 11, 49, fig. 19. Frugoni, *Distant City*, p. 13, notes that the *Liber floridus* example depicts Paradise, but 'has nonetheless drawn upon the theme of the Heavenly Jerusalem as its model'.

11 See B. Kühnel, 'Geography and Geometry of Jerusalem', in *City of the Great King*, ed. Rosovsky, pp. 288–332.

12 Ousterhout, 'Flexible Geography and Transportable Topography', p. 397.

13 Laveden, *Représentation des Villes*, p. 12, fig.1.

14 For the Trinity manuscript, see Brieger, *Trinity College Apocalypse*, p. 49, and for the Valenciennes manuscript, see Gousset, 'Représentation de la Jérusalem', pp. 48–50, n. 8.

15 Frugoni, *Distant City*, p. 21; Gousset, 'Représentation de la Jérusalem',

p. 54: 'quant, au cercle, il est l'image du cosmos'; see also Laveden, *Représentation des Villes*, p. 12: 'Il est intéressant de voir l'image de Jérusalem se modeler sur celle de l'univers ou de ces perfections ultra-terrestres'.

16 Gousset, 'Représentation de la Jérusalem', p. 54; Frugoni, *Distant City*, pp. 22–3; see also Kühnel, 'Use and Abuse of Jerusalem', p. xxiii.

17 Two such twelfth-century examples are British Library MS Add. 32343 (reproduced in D. H. Weiss, '*Hec est Domus Domini firmiter edificata*: The Image of the Temple in Crusader Art', in *Real and Ideal Jerusalem*, ed. Kühnel, pp. 210–17, at p. 213, fig. 3), and Copenhagen MS 736 I (reproduced in Harvey, 'Local and Regional Cartography', p. 475, figure 20.9). The British Library map has five gates in all, one each at the west, south and east, and then two others placed equidistant along the circumference of the walls between the western and eastern gates; whereas the Copenhagen map has four gates, one each at the west, south and east, and one just to the west of north. This arrangement of gates is close to how where gates of the medieval city were located, the Gate of David being to the south, the Gate of St Stephen to the west, and Gate of Tekoa to the east, with the Gate of Benjamin to the north-west (offset by the Temple), the circuit of walls roughly quadrangular in form. The image of the earthly Jerusalem was not quite as perfect as the celestial city, therefore, and yet also not quite the same as the city itself.

18 Alexander, '"Jerusalem the Golden"', p. 260, fig. 7. The positions of the gates follow the same order as the contemporary British Library map.

19 Ibid., p. 256, fig. 2.

20 See H. Rosenau, *The Ideal City* (London, 1983).

21 Alexander, '"Jerusalem the Golden"', p. 264.

22 Of course, these are the cities where the maps now reside, but it seems likely that they were brought there, perhaps by travelling clergy, pilgrims or crusaders, or drawn there, by resident scribes – each of their biographies being little known. See Kühnel, *From the Earthly to the Heavenly Jerusalem*.

23 Little of this genre exists but see Laveden, *Représentation des Villes*; Harvey, 'Local and Regional Cartography'; Frugoni, *Distant City*.

24 This is developed in chapter Two below.

25 Harvey, 'Local and Regional Cartography', p. 493; see also E. Ralph, 'Bristol, *circa* 1480', in *Local Maps and Plans from Medieval England*, ed. R. A. Skelton and P.D.A. Harvey (Oxford, 1986), pp. 309–16.

26 Ralph, 'Bristol', p. 313. The illustration accompanied Ricart's description of Bristol's legendary foundation.

27 On Bristol's early medieval topography see 'Bristol' in *The Atlas of Historic Towns: Bristol, Cambridge, Coventry, Norwich*, ed. M. D. Lobel (London, 1975).

28 Reproduced in Harvey, 'Local and Regional Cartography', fig. 20.9.

29 Reproduced in ibid., p. 474, fig. 20.8, who points out its similarity to Italian plans of cities of this date. For the Copenhagen map see Harvey, 'Local and Regional Cartography', p. 475, fig. 20.9 (he does not draw parallels between this and the Vienna map), and for the Florence map see Laveden, *Représentation des Villes*, p. 32, fig. 2. On other examples see J. Schulz, 'Jacopo de' Barbari's View of Venice: Map Making, CityViews and Moralized Geography before the Year

1500', *Art Bulletin*, LX (1978), pp. 425–74, at p. 456.

30 Reproduced in Harvey, 'Local and Regional Cartography', p. 492, fig. 20.27 (the symbolism of this plan is not addressed by Harvey).

31 S. Reynolds, *Kingdoms and Communities in Western Europe, 900–1300* (Oxford, 1984).

32 On English borough seals see G. Pedrick, *Borough Seals of the Gothic Period* (London, 1904); on Italian communal seals see G. C. Bascapé, *Sigillografia. Il sigillo nella diplomatica, nel diritto, nella storia, nell' arte* (Milan, 1969); for a pan-European comparison see H. Drös und H. Jakobs, 'Die Zeichen einer neuen Klasse, zur Typologie der frühen Stadtsiegel', in *Bild und Geschichte. Studien zur politischen Ikonographie*, ed. K. Krimm and H. John (Sigmaringen, 1997), pp. 125–78.

33 On Oxford see H. E. Salter, *Medieval Oxford* (Oxford, 1936), and also M. D. Lobel, 'Notes on the History of Oxford', *Oxoniensia*, III (1938), pp. 83–102.

34 R.H.C. Davis, 'An Oxford Charter of 1191 and the Beginnings of Municipal Freedom', *Oxoniensia*, XXXIII (1968), pp. 55–65, at 55 and 63, plates 6 and 7.

35 Schulz, 'Jacopo de' Barbari's View of Venice', p. 456, n. 106.

36 Pedrick, *Borough Seals*, for a sample of English seals dating from the thirteenth to fifteenth centuries: a circle of walls appears not only on Oxford's seal of 1191, but also the seals of York (plate 5), Rochester (plate 9), London (plate 11), Worcester (plate 14), Bristol (plate 21), Warwick (plate 26), Rye (plate 35), Wallingford (plate 38), and Shrewsbury (plate 41), though it not always clear if it is a city's defences that are being shown or those of a gateway or castle keep. The seal of the borough of Wilton (Wiltshire), of fourteenth-century date, for example, 'follows in shape the pointed oval which usually distinguishes ecclesiastical seals', see Pedrick, *Borough Seals*, p. 127, plate 31. On ancient and medieval coinage as symbolic of a spherical world, and the orb as an imperial icon, see Woodward, 'Medieval *mappaemundi*', p. 337.

37 For Hildegard of Bingen's 'map' of the cosmos in her *Scivias*, and eleventh- and twelfth-century Neoplatonic discussions of an egg-shaped universe, see P. Dronke, *Fabula: Explorations into the Uses of Myth in Medieval Platonism* (Leiden, 1974), pp. 79–99 (and frontispiece). More generally on mandorla-shaped world maps of the fourteenth and fifteenth centuries, see Woodward, 'Medieval *mappaemundi*', pp. 312–14, fig. 18.22, who points out the likeness between the map's shape and that of the ark; for the oval-shaped map of Rome 'that derives seemingly from illustrations in expanded, late thirteenth-century redactions of the Roman *Mirabilia*', but which appears in manuscripts of Fra Paolino's 'world history' of the 1320s, see Schulz, 'Jacopo de' Barbari's View of Venice', p. 456 and fig. 19.

38 Schulz, "Jacopo de' Barbari's View of Venice', p. 448, who draws a parallel between the image of the world map and the image of the city.

39 J. K. Hyde, 'Medieval Descriptions of Cities', *Bulletin of the John Rylands Library*, XLVIII (1966), pp. 308–40.

40 M. V. Taylor, ed., *Liber Luciani de Laude Cestrie*, Lancashire and Cheshire Record Society, LXIV (1912).

41 English trans. of Latin text of *De Laude Cestrie* in D. M. Palliser, *Chester: Contemporary Descriptions by Residents and Visitors* (Chester, 1980), pp. 6–7.

42 A. Gransden, 'Realistic Observation in Twelfth-century England', *Speculum*, XLVII (1972), pp. 29–51, especially pp. 29, 47.

43 Ibid., p. 47.
44 *De Laude Cestrie* cited in Palliser, *Chester*, pp. 6–7. See also C.A.M. Clarke, *Literary Landscapes and the Idea of England, 700–1400* (Cambridge, 2006), pp. 98–105.
45 On Roman and early medieval Chester, see C. P. Lewis and A. T. Thacker, ed., *A History of the County of Chester*, vol. v/1, *The City of Chester: General History and Topography* (London, 2003).
46 Revelation 21.
47 *De Laude Cestrie* cited in Palliser, *Chester*, pp. 6–7.
48 On the medieval quadrivium and its teaching see for example, P. Kibre, *Studies in Medieval Science: Alchemy, Astrology, Mathematics and Medicine* (London, 1984), pp. 175–91.
49 T. Dean, ed., *The Towns of Italy in the Later Middle Ages* (Manchester, 2000), pp. 16–17.
50 Hyde, 'Medieval Descriptions', p. 332.
51 For Augustine's 'city', see below, note 116.
52 Da Nono cited in Dean, ed., *Towns of Italy*, p. 17.
53 On late medieval Padua see B. G. Kohl, *Padua under the Carrara, 1318–1405* (London, 1998).
54 Hyde, 'Medieval Descriptions', p. 332.
55 As well as those images that showed cities such as Bristol and Vienna with similar conventions, all, it seems, having a shared cosmological symbolism.
56 See Hyde, 'Medieval Descriptions', p. 311, who remarks 'the history of medieval descriptions of cities begins in the second quarter of the eighth century'.
57 B. Krush and W. Levison, ed., *Gregory of Tours, Historiarum libri X*, in *Gregorii episcopi turnonsis Historiarum libri X*, Germaniae Historica, Scriptores rerum Merovingicarum, I.I (Hanover, 1951).
58 Cited in B. S. Bachrach, 'Imperial Walled Cities in the West: An Examination of their Early Medieval Nachleben', in *City Walls: The Urban Enceinte in Global Perspective*, ed. J. D. Tracy (Cambridge, 2000), pp. 192–218, at pp. 200–1.
59 W. H. Stahl, *Roman Science: Origins, Development and Influence to the Later Middle Ages* (Madison, 1962), p. 218.
60 See P. Grierson, 'The European Heritage', in *Ancient Cosmologies*, ed. C. Blacker and M. Loewe (London, 1975), pp. 225–57.
61 G. Lee, ed. and trans., *Plato, Timaeus and Critias* (London, 1977), p. 45.
62 Stahl, *Roman Science*, pp. 193–249; also M. D. Chenu, *Nature, Man, Society in the Twelfth Century: Essays on New Theological Perspectives in the Latin West*, ed. and trans. J. Taylor and L. K. Little (Chicago, IL, 1968), pp. 49–98; M. Gibson, 'The Study of the *Timaeus* in the Eleventh and Twelfth Centuries', *Pensiamento*, XXV (1969); T. Gregory, 'The Platonic Inheritance', in *History of Twelfth-Century Western Philosophy*, ed. P. Dronke (Cambridge, 1988), pp. 54–80; P. E. Dutton, ed., *The Glosae super Platonem of Bernard of Chartres* (Toronto, 1991), pp. 1–136.
63 See W. H. Stahl, trans. *Macrobius' Commentary on the Dream of Scipio* (New York, 1952), pp. 103, 155–56; E. Grant, *Physical Science in the Middle Ages* (Cambridge, 1971), pp. 61, 71–3, quote at p. 73.
64 See Grierson, 'European Heritage'; also E. Grant, *Planets, Stars, and Orbs: the Medieval Cosmos, 1200–1687* (Cambridge, 1994).
65 Stahl, trans., *Macrobius' Commentary*, p. 175.
66 J. H. Waszink, ed., *Timaeus a Calcidio translatus commentarioque instructus*, second edn (London, 1975), p. 149.

67 Stahl, trans., *Macrobius' Commentary*, p. 109.
68 Woodward, 'Medieval *mappaemundi*'.
69 See ibid., pp. 318, 320–21, 343–58.
70 Ibid., p. 319.
71 G.H.T. Kimble, *Geography in the Middle Ages* (London, 1938), pp. 23–39.
72 Woodward, 'Medieval *mappaemundi*', pp. 301–2.
73 J. T. Lanman, 'The Religious Symbolism of the T in T-O Maps', *Cartographica*, XVIII (1981), pp. 18–22.
74 Kühnel, 'Use and Abuse of Jerusalem', pp. xxiii–xxiv.
75 Woodward, 'Medieval *mappaemundi*', p. 335, who refers to Hugh of St Victor's account of the making of a *mappamundi* and his likening the shape of the earth to the ark.
76 B. Obrist, 'Wind Diagrams and Medieval Cosmology', *Speculum*, LXXII (1997), pp. 58–65.
77 Matthew 24:31; Revelation 7:1; J.L.E. Dreyer, *History of the Planetary Systems from Thales to Kepler* (Cambridge, 1906), pp. 214–19; cf. Woodward, 'Medieval *mappaemundi*', p. 319.
78 Obrist, 'Wind Diagrams and Medieval Cosmology', p. 58, figs 17 and 18.
79 Ibid., p. 63, fig. 19.
80 Woodward, 'Medieval *mappaemundi*', p. 319; Obrist, 'Wind Diagrams and Medieval Cosmology', p. 64, n. 110.
81 Lee, trans., *Plato, Timaeus*, p. 49.
82 See J. J. Sheridan, trans. *Alan of Lille: The Plaint of Nature* (Toronto, 1980); also B. Stock, *Myth and Science in the Twelfth Century: A Study of Bernard Silvester* (Princeton, NJ, 1972).
83 See J. Taylor, trans., *The Didascalicon of Hugh of St Victor: A Medieval Guide to the Arts* (New York, 1961), p. 25.
84 See ibid., pp. 64–7, 178–9.
85 Lee, trans. *Plato, Timaeus*, p. 44; Obrist, 'Wind Diagrams and Medieval Cosmology', p. 64.
86 Woodward, 'Medieval *mappaemundi*', p. 335, also p. 336, fig. 18.38.
87 Cited in Gousset, 'Représentation de la Jérusalem', p. 54, note 26. English translation from Frugoni, *Distant City*, p. 21.
88 Woodward, 'Medieval *mappaemundi*', p. 335; U. Eco, *The Aesthetics of Thomas Aquinas*, trans. H. Bredin (Cambridge, MA, 1988), p. 79; see Frugoni, *Distant City*, p. 21.
89 A. J. Gurevich, *Categories of Medieval Culture*, trans. G. L. Campbell (London, 1985), pp. 58–61; Stock, *Myth and Science*, p. 161.
90 Cited in Eco, *Aesthetics of Thomas Aquinas*, p. 77.
91 See Gibson, 'Study of the *Timaeus*'; R. W. Southern, *Platonism, Scholastic Method and the School of Chartres* (Reading, 1978).
92 Kühnel, 'Use and Abuse of Jerusalem', p. xxiii.
93 Ibid., p. xxiv.
94 Woodward, 'Medieval *mappaemundi*', pp. 515, 514; see also Kühnel, 'Use and Abuse of Jerusalem', pp. xxiii–xxiv; Kühnel, 'Geography and Geometry of Jerusalem', p. 319.
95 Whether time was linear or cyclical, and whether the world was created (as Plato held) or eternal (as Aristotle held), were issues debated in Classical times, and resonated throughout the Middle Ages, too: see G.E.R. Lloyd, 'Greek Cosmologies', in *Ancient Cosmologies*, ed. Blacker and Loewe, pp. 206–7; Grierson, 'European Heritage', pp. 236–47.
96 Isidore wrote: 'The circle of lands [*orbis*] is so called from its roundness, which is like that of a wheel', cited in Woodward, 'Medieval *mappae-*

mundi', p. 320. *Annus* meant 'ring', Latin for year. See also Obrist, 'Wind Diagrams and Medieval Cosmology', p. 63, figs 19 and 20.

97 Woodward, 'Medieval *mappaemundi*', p. 339, citing E. Kitzinger, 'World Map and Fortune's Wheel: A Medieval Mosaic Floor in Turin', *Proceedings of the American Philosophical Society*, CXVII (1973), pp. 344–73.

98 Dronke, *Fabula*, p. 137. See also Sheridan, trans., *Alan of Lille*; Stock, *Myth and Science*.

99 O. Limor, 'The Place of the End of Days: Eschatological Geography in Jerusalem', in *Ideal and Real Jerusalem*, ed. Kühnel, pp. 13–22; the cog-shape of Jerusalem is most apparent in images such as the twelfth-century Copenhagen and Florence maps of the city, see Harvey, 'Local and Regional Cartography', p. 475, fig. 20.9 and Laveden, *Représentation des Villes*, pp. 12–13, 32, fig. 2, and see also Schulz, 'Jacapo de' Barbari's View of Venice', p. 457, fig. 18; on the *machina mundi* see L. Thorndike, ed. and trans. *The 'Sphere' of Sacrobosco and its Commentators* (Chicago, IL, 1949), p. 12, where he refers to the use of this 'Lucretian expression' by both Robert Grosseteste and John of Sacrobosco in their cosmographies of the thirteenth century. Sacrobosco thus writes: 'the machine of the universe is divided into two . . .' (Thorndike, ed. and trans., *Sphere of Sacrobosco*, p. 119). Lucretius' writings were also known to Isidore of Seville (see Woodward, 'Medieval *mappaemundi*', p. 301), and in his *Consolation of Philosophy*, Boethius refers to 'the great machine/ which unity maintains/with motions beautiful' in the poetic ending of Book Two, having discussed earlier the work of Fortune (see V. Watts, ed. and trans. *Boethius, The Consolation of Philosophy* (London, 1999), p. 45).

100 Bettenson, ed. and trans., *Augustine, City of God*, p. 600: 'The first founder of the earthly city was, as we have seen, a fratricide; for, overcome by envy, he [Cain] slew his own brother, a citizen of the Eternal City [i.e., *The City of God*], on pilgrimage in this world. Hence, it is no wonder that long afterwards this first precedent . . . was answered by a kind of reflection, by an event of the same kind at the founding of the city which was to be the capital of the earthly city of which we are speaking [Rome], and was to rule over so many peoples . . . For this is how Rome was founded, when Remus, as Roman history witnesses, was slain by his brother Romulus . . .'; and in Book 17, 3 (p. 714): 'The fact that Jerusalem is called The City of God has a double reference, combined as it is with the prophecy of the future house of God in that city. This prophecy seems to have its fulfilment when King Solomon builds that renowned Temple. But this was not only an event in the history of the earthly Jerusalem; it was also a symbol of the Jerusalem in heaven'; and Book 15, 2 (p. 597): 'One part of the earthly city has been made into an image of the Heavenly City, by symbolising something other than itself, namely that other City; and for that reason it is a servant. For it was established not for its own sake but in order to symbolise another City, and since it was signified by an antecedent symbol, the foreshadowing symbol was itself foreshadowed . . . Thus we find in the earthly city a double significance: in one respect it displays its own presence, and in the other it serves by its own presence to signify the Heavenly City'.

101 Ibid., pp. 471, 429.

102 Rosenau, *Ideal City*, p. 28, fig. 14, refers to the four walls shown on this image (which comes from Hildegard of Bingen's *Scivias*) as a 'building', but really it shows more similarity to quadrangular forms of the Heavenly

Jerusalem, with its walls 'four-square' as Revelation dictates. On the symbolism of the corner stone see G. B. Ladner, 'The symbolism of the Biblical Corner Stone in the Mediaeval West', *Mediaeval Studies*, IV (1942), pp. 43–60.

103 Lee, trans., *Plato, Timaeus*, p. 50; Bettenson, ed. and trans., *Augustine, City of God*, pp. 453–56; see also Lloyd, 'Greek Cosmologies', pp. 198–224, at pp. 202–3.

104 J. B. Friedman, 'The Architect's Compass in Creation Miniatures of the Later Middle Ages', *Traditio*, XXX (1974), pp. 419–29, at pp. 424, 420, fig. 11. Friedman found 38 such scenes in manuscripts of 1000–1400, and lists them (p. 420, n. 2).

105 Ibid., pp. 422, 424.

106 See ibid., p. 423, figs 3–10. On architects' use of measuring instruments in design work see J. Gimpel, *The Cathedral Builders*, trans. T. Waugh (London, 1983), pp. 89–121; also L. R. Shelby, 'Medieval Masons' Tools II. Compass and Square', *Technology and Culture*, VI (1965), pp. 236–48; L. R. Shelby, 'The Geometrical Knowledge of the Mediaeval Master Masons', *Speculum*, XLVII (1972), pp. 395–421. Shelby does not consider the symbolic attributes of these instruments or their geometry.

107 See also chapters Three and Four. See E. A. Zaitsev, 'The Meaning of Early Medieval Geometry, from Euclid and Surveyors' Manuals to Christian Philosophy', *Isis*, XC (1999), pp. 522–53. For example, the cosmographies of Macrobius and Calcidius with their calculations of the size of the universe and its extent, illustrated with diagrams of the circular-shaped cosmos – see Stahl, trans., *Macrobius' Commentary*, pp. 174–81 and Waszink, ed., *Timaeus*, pp. 120–23, 148–50. On the uses of geometry in cosmimetry, see, for example, Hugh of St Victor's *Practica Geometriæ* of the early twelfth century, where he explains the need to reduce the sphere of the world 'to model form' and devotes the third part of his work to the 'measurement of the cosmos' and describes 'how to measure it', F. A. Homann, trans., *Practical Geometry [Practica Geometriae] attributed to Hugh of St Victor* (Milwaukee, WI, 1991), pp. 35, 57.

108 Quoted in Friedman, 'Architect's Compass in Creation Miniatures', p. 425, citing F. H. Colson and G. H. Whitaker, trans., *Philo* (Cambridge, MA, 1929), I, pp. 17–20. The question remains as to how much of Philo's writings were known and read in the medieval Latin West – see Grierson. 'European Heritage', p. 229.

109 Architects with compass and a set-square are depicted in contemporary illustrations, such as the thirteenth-century *Vie de Saint Offar* (see Gimpel, *Cathedral Builders*, p. 44), the image of Eudes de Montreuil, master mason of Paris (fl. *c.*1220–89), carved onto his tomb (in W. R. Lethaby, *Mediæval Art: From the Peace of the Church to the Eve of the Renaissance 312–1350* (London, 1912), pp. 251–4, fig. 110), and a thirteenth-century miniature in Dublin Trinity College Library MS. E.1.40 (see Shelby, 'Medieval Masons' Tools', fig. 2).

110 S. K. Victor, ed. and trans., *Practical Geometry in the High Middle Ages: 'Artis Cuiuslibet Consummatio' and the 'Pratike de Geometrie'* (Philadelphia, PA, 1979), pp. 32–7, at p. 37. See also Ladner, 'Symbolism of the Biblical Corner Stone'.

111 See also chapter Four. On the compass and squares used by medieval masons and architects, see Shelby, 'Medieval Masons' Tools'; for more detailed discussions of the quadrant, the 'geometric square', and the astrolabe and their use, see R. T. Gunther, *Early Science in Oxford*, vol. I

(Oxford, 1923), pp. 333–44, and E. R. Kiely, *Surveying Instruments: Their History and Classroom Use* (New York, 1947), pp. 64–83. On medieval accounts of their use see, for example, Hugh of St Victor's *Practica Geometriae* where he describes the use of the astrolabe in surveying work in Homann, trans., *Practical Geometry*, pp. 40–47, and also the anonymous 'practical geometry' written in northern France known by the incipit, *Artis cuiuslibet consummatio*, which also describes using instruments for measurement in planimetry, altimetry and cosmimetry: see Victor, ed. and trans., *Practical Geometry in the High Middle Ages*. A thirteenth-century manuscript (New York Public Library De Ricci #69) of John of Sacrobosco's *Tractatus de sphera*, a cosmography that 'welded together Macrobius and Ptolemy and frosted it over with Alfraganus' – opens 'with an illuminated initial letter showing a man seated and holding an astrolabe' (Thorndike, ed. and trans., *Sphere of Sacrobosco*, pp. 21, 69). See also the illustrations showing survey work in Gunther, *Early Science*, p. 334, and plate opposite p. 335, and the well-known example in Villard de Honnecourt's *Sketchbook* showing how 'one measures the height of a tower', in F. Bucher, *Architector: The Lodgebooks and Sketchbooks of Medieval Architects*, vol. I (New York, 1979), p. 124, plate 40.

TWO: URBAN FORMS

1 For an overview of these studies see J.W.R. Whitehand, 'Background to the Urban Morphogenetic Tradition', in *The Urban Landscape: Historical Development and Management: Papers by M.R.G. Conzen*, ed. J.W.R. Whitehand (London, 1981), pp. 1–24; K. D. Lilley, 'Reading the Medieval Urban Landscape: Approaches to the Morphological Study of Landscape', in *Reading the Landscape*, ed. D. Munro (Edinburgh, 2002), pp. 70–82.

2 K. D. Lilley, 'Mapping the Medieval City: Plan Analysis and Urban History', *Urban History*, XXVII (2000), pp. 5–30.

3 M. W. Beresford, *New Towns of the Middle Ages: Town Plantation in England, Wales and Gascony* (London, 1967); C. Meckseper, *Kleine Kunstgeschichte der Deutschen Stadt im Mittelalter* (Darmstadt, 1982); F. Divorne, B. Gendre, B. Lavergne and P. Panerai, *Les Bastides d'Aquitaine, du Bas-Languedoc et du Béarn, Essai sur la Regularite* (Brussels, 1985); D. Friedman, *Florentine New Towns: Urban Design in the Late Middle Ages* (Cambridge, MA, 1988); A. Lauret, R. Malebranche and G. Séraphin, *Bastides, Villes Nouvelles du Moyen Age* (Toulouse, 1988).

4 See K. D. Lilley, *Urban Life in the Middle Ages, 1000–1450* (London, 2002), pp. 75–93, 106–36. Not all new towns had 'regular forms' and not all 'regular forms' are associated with new towns. Indeed, medieval town planning often gave rise to various urban forms.

5 See K. D. Lilley, 'Urban Landscapes and their Design: Creating Town from Country in the Middle Ages', in *Town and Country in the Middle Ages*, ed. K. Giles and C. Dyer (Leeds, 2005), pp. 229–49.

6 See T. Reuter, ed., *Alfred the Great: Papers from the Eleventh-Centenary Conferences* (Aldershot, 2003), especially the papers by David Hill and Derek Keene.

7 M. Biddle and D. Hill, 'Late Saxon Planned Towns', *Antiquaries Journal*, LI(1971), pp. 70–85.

8 Ibid.; M. Aston and C. J. Bond, *The Landscape of Towns* (London, 1976);

J. Haslam, ed., *Anglo-Saxon Towns in Southern England* (Chichester, 1984).

9 Asser in his *Life of Alfred* wrote: 'I am speaking here of fortifications commanded by the king which have not yet been begun, or else, having been begun late in the day, have not been brought into completion', S. Keynes and M. Lapidge, ed. and trans., *Alfred the Great, Asser's 'Life of Alfred' and Other Contemporary Sources* (London, 1983), p. 102.

10 J. Munby, 'Saxon Chichester and its Predecessors', in *Anglo-Saxon Towns*, ed. Haslam, pp. 315–30.

11 P. Crummy, 'The System of Measurement Used in Town Planning from the Ninth to the Thirteenth Centuries', in *Anglo-Saxon Studies in Archaeology and History*, I, ed. S. C. Hawkes, D. Brown and J. Campbell (Oxford, 1979), pp. 149–64; J. Haslam, 'The Metrology of Anglo-Saxon Cricklade', *Medieval Archaeology*, XXX (1986), pp. 99–102.

12 Crummy, 'System of Measurement', p. 153, notes, 'subject to appropriate detailed study of their street layouts, a distinctive 'Wessex-style' plan based on 16-pole units can be postulated at the southern towns of Chichester, Wareham and Wallingford, as well as Winchester'. Haslam, 'Anglo-Saxon Cricklade', would seem to confirm this postulation.

13 R. A. Hall, 'The Five Boroughs of the Danelaw: A Review of Present Knowledge', *Anglo-Saxon England*, XVIII (1989), pp. 149–206.

14 Translation from Biddle and Hill, 'Late Saxon Planned Towns', p. 83, citing W. H. Stevenson, ed., *Asser's Life of Alfred* (Oxford, 1904), p. 77. For a different translation of this same passage see Keynes and Lapidge, ed. and trans., *Alfred the Great*, p. 101: 'and what of the cities and towns to be rebuilt and of others to be constructed where previously there were none?' Also, Asser writes: 'I am speaking here of fortifications commanded by the king which have not yet been begun, or else, having been begun late in the day, have not been brought into completion' (p. 102).

15 R. Hodges, *Towns and Trade in the Age of Charlemagne* (London, 2000). Hodges notes the connections between Alfred's England and Carolingian Europe, and suggests that Charlemagne might have been familiar with 'the ethos of classical urbanism' that existed in the Levant, and concedes that 'it is difficult to imagine that these places did nor have an influence' on the 'urban rebirth of north-west Europe' (p. 101). In particular Hodges draws attention to the new town of 'Anjar of the Umayyad dynasty, with its highly formalized quadrangular layout. This seems to have been laid out in imitation of the Roman *castrum* layout and might have been the kind of place that nobles, pilgrims, scholars and traders from Carolingian Europe saw on their travels to the Levant and brought back with them: for a detailed discussion of the plan of 'Anjar, see R. Hillenbrand, 'Anjar and Early Medieval Urbanism', in *The Idea and Ideal of the Town between Late Antiquity and the Early Middle Ages*, ed. G. P. Brogiolo and B. Ward-Perkins (Leiden, 1999), pp. 59–98. Hodges, *Towns and Trade*, pp. 107–17, 123–4, mostly alludes to connections in 'town-making' between Charlemagne, Alfred and the eastern Mediterranean, rather than spelling these out in detail. For a Roman model closer to home one only has to turn to the plan of Portchester to see parallels with the square plans of the new *burhs*, see C. A. Raleigh Radford, 'The Later Pre-Conquest Boroughs and their Defences', *Medieval Archaeology*, XIV (1970), pp. 83–103, at p. 100. Portchester was one of the places listed in the Burghal Hidage documents of the early tenth century which includes Alfred's Wessex burhs, see D. Hill, 'The

Burghal Hidage: The Establishment of a Text', *Medieval Archaeology*, XIII (1969), pp. 84–92.

16 For example, in ninth-century Apocalypse manuscripts showing the heavenly Jersusalem produced in Carolingian cities such as Valenciennes and Cambrai. See M-T. Gousset, 'La représentation de la Jérusalem céleste à l'époque Carolingienne', *Cahiers Archéologiques*, XXIII (1974), p. 60. On ninth-century urban development in the Rhine, Somme and Meuse valleys see A. Verhulst, *The Rise of Cities in North-West Europe* (Cambridge, 1999), pp. 44–76. Alfred visited Carolingian France as a boy, as well as Rome, and 'resolved to revive learning in his own kingdom' in later years 'and may have made a special effort to ascertain in particular how the Carolingian kings had dealt with the Viking raiders, when confronted with the same problem himself', Keynes and Lapidge, ed. and trans., *Alfred the Great*, p. 14.

17 On Carolingian Aachen see M. Schmitt, *Aachen, Deutscher Städtatlas 4* (Altenbeken, 1984). For a discussion of Alfred's London and its restoration see Hodges, *Towns and Trade*, pp. 112–17, drawing upon J. Clark, 'King Alfred's London and London's King Alfred', *London Archaeologist*, IX (1999), pp. 35–8. See also T. Tatton-Brown, 'The Topography of Anglo-Saxon London', *Antiquity*, LX (1986), pp. 21–8, especially pp. 25–6; A. Vince, *Saxon London: An Archaeological Investigation* (London, 1990). After taking London from the Danes in 886, Alfred 'refurbished the walls, repopulated the city, and during a conference to discuss its restoration (*instauracio*) assigned to various magnates plots of land bounded by streets': Biddle and Hill, 'Late Saxon planned towns', p. 83. The areas of Aachen and London that look similar in form are those streets and plots that lie just to the north of Charlemagne's palace at Aachen and the area around Eastcheap in London. See the plans in Schmitt, *Aachen*, and Tatton-Brown, 'Anglo-Saxon London', the latter noting 'it was Alfred who repopulated the whole of the southern part of the eastern city by laying out a new 'planned' town in the area' (p. 25).

18 As noted by A. F. Payne, *King Alfred and Boethius: An Analysis of the OE Version of the 'Consolation of Philosophy'* (Madison, WI, 1968), pp. 60–61: 'in the Latin text [of Boethius], the "country" is the country of the philosopher, the city state of the mind; in the OE text [of Alfred], the country is the heavenly city of all those who desire the right: . . . "you are one of those who are righteous and will the right; these are the inhabitants of the Heavenly Jerusalem"', citing W. Sedgefield, ed. and trans., *King Alfred's Old English Version of Boethius' 'De Consolatione Philosophiae'* (Oxford, 1899), p. 11. Cf. V. Watts, ed. and trans. *Boethius, The Consolation of Philosophy* (London, 1999), pp. 16–18.

19 Watts, ed. and trans., *Boethius, Consolation*, pp. 45, 66–7, 72: 'all this chain of things/ in earth and sea and sky/ one ruler holds in hand'; the microcosmicism in *Consolation* derives from Neoplatonic sources, including Calcidius and Macrobius. Alfred's translation of Boethius' text and his rendering of Boethian ideas of order are discussed in detail by Payne, *Alfred and Boethius*; see especially pp. 22–3, 25–6, 44, 68, 75, 86, 97, 129 for his treatment of Neoplatonisms on the hierarchy and order of things, on kingship and on the body. Alfred amplified aspects of the Latin text. He seems to be thinking of his own place in the world when discussing the three orders, a Neoplatonic convention (see P. E. Dutton, '*Illustre civitatis et populi exemplum*: Plato's *Timaeus* and the transmission of Calcidius to the end of the twelfth century of a

tripartite scheme of society', *Mediaeval Studies*, XLV (1983), pp. 79–119): 'in the case of the king, the resources and tools with which to rule are that have his land fully manned: he must have praying men, fighting men and working men', Keynes and Lapidge, ed. and trans. *Alfred the Great*, p. 132. He also states: '"Look, Wisdom, you know that desire for and possession of earthly power never pleased me overmuch, and that I did not unduly desire this earthly rule, but that nevertheless I wished for tools and resources for the task that I was commanded to accomplish, which was that I should virtuously and worthily guide and direct the authority which was entrusted to me"' (p. 132, cf. pp. 40–44). For Alfred's translation of Augustine's *Soliloquies* see Keynes and Lapidge, ed. and trans., *Alfred the Great*, pp. 138–52, especially at p. 140, where the cosmic order and celestial hierarchy are discussed. Writing in the 1140s in his *Cosmographia*, Bernard Silvester believed 'the cosmic order is a kind of continuum', see B. Stock, *Myth and Science in the Twelfth Century: A Study of Bernard Silvester* (Princeton, NJ, 1972), p. 161.

20 In a letter from Fulco, archbishop of Reims, to Alfred, written *c.* 886 at the time when the king was looking to revive learning in England, Fulco wrote: 'you attend to the good of the kingdom divinely entrusted to you, seeking or safeguarding its peace with warlike weapons, and also, with your mind continually on religious matters, you take care to increase the dignity of the ecclesiastical order with spiritual weapons': Keynes and Lapidge, ed. and trans., *Alfred the Great*, pp. 182–86, at p. 182.

21 B. Zientara, 'Socio-economic and Spatial Transformation of Polish Towns during the Period of Location', *Acta Poloniae Historica*, XXXIV (1976), pp. 57–83; M. Prestwich, *Edward I* (New Haven, CT, 1988), pp. 298–311; Beresford, *New Towns*, pp. 28–35; Lauret et al., *Bastides*, pp. 21–39.

22 On dynastic struggles and territorial expansion in Europe during the twelfth and thirteenth centuries see R. Bartlett, *The Making of Europe: Conquest, Colonization and Cultural Change 950–1350* (London, 1993), pp. 24–59.

23 This symbolic significance of the heavenly city was probably in Alfred's mind as he wrote, 'and after the Day of Judgement it is promised to us that we may be able to see God clearly, to see Him exactly as He is, and henceforth to know Him perfectly as He now knows us', in his translation of Augustine's *Soliloquies*: Keynes and Lapidge, ed. and trans., *Alfred the Great*, p. 149. In the sixteenth century the New Jerusalem was invoked by Spanish Franciscans settling and colonizing the Americas, where new towns were also created four-square in plan. See R. Kagan, 'A World Without Walls: City and Town in Colonial Spanish America', in *City Walls. The Urban Enceinte in Global Perspective*, ed. J. D. Tracy (Cambridge, 2000), pp. 117–52.

24 On the '*bastides*' see Beresford, *New Towns*; Divorne et al., *Les Bastides*; Lauret et al., *Bastides*. *Bastide* is derived from the Latin *bastire* and French *bâtir*, 'to build'; see A. Randolph, '*Bastides* of South-west France', *Art Bulletin*, LXXVII (1995), p. 291, n. 11.

25 For example, by A.E.J. Morris, *History of Urban Form: Before the Industrial Revolutions* (Harlow, 1994), pp. 93–8; L. Benevolo, *The European City*, trans. C. Ipsen (Oxford, 1993), pp. 69–70.

26 Beresford, *New Towns*, pp. 142–78.

27 Randolph, '*Bastides* of South-west France', p. 306. He states that his 'essay posits a functional analysis of the grid plan', p. 291.

28 Morris, *History of Urban Form*, p. 119 notes that 'it is a grid of

convenience – the quickest and most equitable way of laying out a town on a new site'.

29 E. Mâle, *Religious Art in France: The Thirteenth Century* (Princeton, NJ, 1984), p. 16.

30 Randolph, '*Bastides* of South-west France', p. 301, uses the analogy of 'a net, thrown upon the site and adapting to its nuances'. Morris, *History of Urban Form*, pp. 119–32, erroneously uses the term '*bastide*' also to describe new towns in general in England and Wales, as well as France.

31 Lauret et al., *Bastides*, pp. 58–78, Divorne et al., *Les bastides*, pp. 40–41.

32 Lauret et al., *Bastides*, p. 72; Divorne et al., *Les Bastides*, pp. 73, 77.

33 Some *bastide* forms are circular in shape, see below.

34 In Moravia, for example, the layout of Moravská Třebová has twelve square-shaped street blocks set around a central square marketplace, encompassed by a squarely arranged circuit of defences, and shows similarities in form with other Moravian towns such as Nový Jiřín, and those in Bohemia such as České Budějovice. Again, all are new towns of thirteenth-century date. See G. Gutkind, ed., *Urban Development in East-Central Europe: Poland, Czechoslovakia and Hungary* (London, 1972), p. 287; Morris, *History of Urban Form*, pp. 149–51; Divorne et al., *Les Bastides*, p. 91. Also M. Koter and M. Kulesza, 'The Plans of Medieval Polish Towns', *Urban Morphology*, III (1999), pp. 63–78; W. Kalinowski, 'City Development in Poland', in *Urban Development in East-Central Europe: Poland, Czechoslovakia and Hungary*, ed. Gutkind, pp. 26–51, especially pp. 26–39; T. Zagrodski, 'Les plans de villes créées au moyen âge en Pologne et la tradition des règles gromaticales', *Revue des Archéologiques et Historiens d'Art de Louvain*, IX (1976), pp. 105–18.

35 F. Bucher, 'Medieval Architectural Design Methods, 800–1560', *Gesta*, XI (1972), pp. 37–43.

36 Randolph, '*Bastides* of South-west France', p. 301.

37 Divorne et al., *Les Bastides*, p. 63.

38 This is borne out by a study of twelfth-century new towns in southern England; see the examples of Portsmouth and Newport, which have plans made up of street-blocks in a parallelogram arrangement but with curved streets and plots; K. D. Lilley, 'Urban Planning and the Design of New Towns in the Middle Ages: The Earls of Devon and their "New Towns"', *Planning Perspectives*, XVI (2001), pp. 1–24. For other similar examples, see R. Hager, *The Zähringer New Towns*, exh. cat., Swiss Federal Institute of Technology, Zurich (1966).

39 Lauret et al., *Bastides*, p. 72.

40 J. L. García Fernández, 'La rica experiencia urbanística del medievo español', in *La Ciudad Hispanoamericana. El Sueño de un Orden*, exh. cat., Ministerio de Obras Públicas y Urbanismo, Madrid (1989), p. 90.

41 The rectangular-shaped street blocks of New Winchelsea (Sussex, England), laid out in the 1280s, were locally referred to as 'quarters' in contemporary written accounts, see Beresford, *New Towns*, p. 18.

42 Deliberate curvature in medieval urban forms is controversially discussed by K. Humpert and M. Schenk, *Entdeckung der mittelalterlichen Stadtplanung. Das Ende vom Mythos der 'gewachsenen Stadt'* (Stuttgart, 2001). See the review by U. Fischer in *Urban Morphology*, VII (2003), pp. 125–7; and B. Scholkmann, 'Zwischen Mythos und Befund. Eine kritische Bilanz zum Thema "Die vermessene Stadt" aus Sicht der archäologischen Stadtkernforschung', *Mitteilungen der Deutschen Gesellschaft für Archäologie des Mittelalters und der Neuzeit* (Paderborn, 2004), pp. 180–84. Those

involved in the 'picturesque' town planning movement of the late nineteenth and early twentieth centuries argued for deliberately created curvatures in medieval town plans, see K. D. Lilley, 'Modern Visions of the Medieval City: Competing Conceptions of Urbanism in European Civic Design', *Environment and Planning b: planning and design*, XXVI (1999), pp. 427–46. There are, of course, differing degrees of curvature in medieval urban forms.

43 The circular-shaped world is discussed by Robertus Anglicanus in his *Commentary on The Sphere of Sacrobosco*: 'a model of the world existent in the divine mind before the creation of the world after whose likeness the world was created. And, since that world was in the divine mind and that which is in God is one with God, as Augustine holds, therefore, as God had neither beginning nor end, so no more had that world. Moreover a figure without beginning or end is circular': L. Thorndike, ed. and trans., *The 'Sphere' of Sacrobosco and its Commentators* (Chicago, IL, 1949), p. 208. For an Aristotelian view that the world had no beginning and no end, and one that is at odds with Scripture, see E. Grant, *Planets, Stars, and Orbs: the Medieval Cosmos, 1200–1687* (Cambridge, 1994), p. 63.

44 See above, chapter One.

45 S. K. Victor, ed. and trans., *Practical Geometry in the High Middle Ages: 'Artis Cuiuslibet Consummatio' and the 'Pratike de Geometrie'* (Philadelphia, PA, 1979), p. 219.

46 H. B. Clarke and A. Simms, 'Towards a Comparative History of Urban Origins', in *The Comparative History of Urban Origins in Non-Roman Europe*, ed. H. B. Clarke and A. Simms (Oxford, 1985), pp. 547–677.

47 L. Swan, 'Monastic Proto-towns in Early Medieval Ireland: The Evidence of Aerial Photography, Plan Analysis and Survey', in *Comparative History of Urban Origins*, ed. Clarke and Simms, pp. 77–102, see fig. 98. They have also been identified in Wales and south-west England. See L. Butler, '"The Monastic City" in Wales: Myth or Reality?', *Bulletin of the Board of Celtic Studies*, XXVIII (1979), pp. 458–67.

48 Swan, 'Monastic Proto-towns', p. 84, figure 4.4.

49 See C. Frugoni, *A Distant City: Images of Urban Experience in the Medieval World*, trans. W. McCuiag (Princeton, NJ, 1991), pp. 20–23.

50 Text translated in C. Doherty, 'The Monastic Town in Early Medieval Ireland', in *Comparative History of Urban Origins*, ed. Clarke and Simms, pp. 45–75, at p. 59.

51 H. J. Lawlor, *Chapters in the Book of Mulling* (Edinburgh, 1897), pp. 167–85. The figure is on p. 167. For a translation of the diagram see Butler, 'Monastic City', p. 459. Further discussion in K. Hughes, *The Church in Early Irish Society* (London, 1966), p. 199.

52 Doherty, 'Monastic Town', p. 57, discusses the celestial city in Ireland, but except in saying that 'the shape of monastic sites shows that there was an attempt to make them conform in some way to the celestial model' (p. 60), he does not draw attention to its imagined *circular* form and the circular forms of the Irish monastic towns.

53 Examples may arise with further study. In the meantime it is worth noting that key monastic towns in England in the eighth to eleventh centuries have either a rectilinear form, as at Bury St Edmunds, or a plan based on a triangular-shaped marketplace outside the monastery gates, as at St Albans and Coventry. The symbolism of the triangle (as a sign of the Trinity) has not gone unnoticed. See T. R. Slater, 'Medieval

'Town-founding on the Estates of the Benedictine Order in England', in *Power, Profit and Urban Land. Landownership in Medieval and Early Modern Northern European Towns*, ed. F.-E. Eliassen and G. A. Ersland (Aldershot, 1996), pp. 70–92; B. Gauthiez, 'The Planning of the Town of Bury St Edmunds: a Probable Norman Origin', in A. Gransden, ed., *Bury St Edmunds: Medieval Art, Architecture, Archaeology and Economy* (London, 1998), pp. 81–97.

54 K. Pawlowski, 'Villes et villages circulaires du Languedoc', *Annales du Midi*, IX (1987), pp. 407–28.

55 Lauret et al., *Bastides*, p. 59.

56 Ibid, p. 288.

57 Harvey, 'Local and Regional Cartography', p. 492, fig. 20.27: the map (Archivio di Stato, Siena, Capitoli 3, fols. 25v–26) shows 'existing (named) and projected (numbered) building plots . . . and so might have been produced to fulfil some planning or administrative need'.

58 J.J.G. Alexander, '"Jerusalem the Golden": Image and Myth in the Middle Ages in Western Europe', in *The Real and Ideal Jerusalem in Jewish, Christian and Islamic Art*, ed. B. Kühnel (Jerusalem, 1998), p. 260, fig. 7 (the 'Crusader map': Hague, Koninklijke Bibliotheek, MS. 76, fol. 5). For the 'Florence map' see R. Ousterhout, 'Flexible Geography and Transportable Topography', in *Real and Ideal Jerusalem*, ed. Kühnel, p. 396, fig. 3. See also the fourteenth-century copy of a *c.* 1180 original map of Jerusalem (Arnamagnæan Commission, Copenhagen, MS. 736 I, 4to, fol. 2r): 'within the circular diagrammatic wall are shown two main thoroughfares in the form of a cross', Harvey, 'Local and Regional Cartography', p. 475, fig. 20.9.

59 E. Ralph, 'Bristol, *circa* 1480', in *Local Maps and Plans from Medieval England*, ed. R. A. Skelton and P.D.A. Harvey (Oxford, 1986), p. 313, where the similarity between the 'map' of the centre of Bristol and the configuration of its layout is noted. The suggestion is that its creator 'was more interested in the pattern of the design than in the cartographic accuracy of this plan' (p. 314). The similarity between it and images of the heavenly city, and its possible symbolism, is not noted.

60 For each town see Lauret et al., *Bastides*, pp. 72, 250–51, 286, 288. The *bastide* of Valentine also has a curvilinear defensive circuit and squared internal layout, see p. 303.

61 For Nubrandenburg, see Morris, *History of Urban Form*, p. 137, figure 4.69; for Strzelce Krajeńskie, see Koter and Kulesza, 'Medieval Polish Towns', p. 71, figure 10; for Cittadella, see Ousterhout, 'Flexible Geography and Transportable Topography', p. 402, fig. 7.

62 Friedman, *Florentine New Towns*, pp. 122–9.

63 A cosmographical matter that bothered Rabanus Maurus in trying to reconcile the Neoplatonic and Biblical conceptions of the shape of the world – a circle and yet a square (see chapter One).

64 U. Eco, *The Aesthetics of Thomas Aquinas*, trans. H. Bredin (Cambridge, MA, 1988), p. 90. On harmony and microcosmicism see R. Allers, 'Microcosmus: From Anaximandros to Paracelsus', *Traditio*, II (1944), pp. 375–9.

65 See Grant, *Planets, Stars, and Orbs*.

66 J. Taylor, trans., *The Didascalicon of Hugh of St Victor: A Medieval Guide to the Arts* (New York, 1961), p. 69.

67 W. H. Stahl, *Roman Science: Origins, Development and Influence to the Later Middle Ages* (Madison, 1962), p. 200.

68 See chapter One. U. Eco, *Art and Beauty in the Middle Ages* (London,

1986), pp. 35–6, discusses this with respect to the idea of *homo quadratus*, 'squared man'.

69 See Stahl, *Roman Science*, eg. pp. 16–17, 195–203, also chapter Two.

70 See Eco, *Art and Beauty*, pp. 28–51; Eco, *Aesthetics of Thomas Aquinas*, pp. 71–98. See also Mâle, *Religious Art*, pp. 31–64; E. Panofsky, *Gothic Architecture and Scholasticism* (Latrobe, PA, 1951), pp. 20–35.

71 A. J. Gurevich, *Categories of Medieval Culture*, trans. G. L. Campbell (London, 1985), p. 61.

72 Eco, *Art and Beauty*, p. 116.

73 Mâle, *Religious Art*.

74 Haslam, 'Anglo-Saxon Cricklade', fig. 6.

75 See the plans of these in Biddle and Hill, 'Late Saxon Planned Towns', fig. 4; Aston and Bond, *Landscape of Towns*, fig. 11.

76 Haslam, 'Anglo-Saxon Cricklade', p.100.

77 See K. D. Lilley, 'Taking Measures Across the Medieval Landscape: Aspects of Urban Design Before the Renaissance', *Urban Morphology*, II (1998), pp. 82–92.

78 See Gutkind, ed., *Urban Development in East-Central Europe*, p. 287; Morris, *History of Urban Form*, pp. 149–55; Divorne et al., *Les Bastides*, p. 91.

79 For example in the Dordogne, the plan of Monpazier is highly orthogonal, a new town created *c.* 1284 for the English crown; Molières likewise was established *c.* 1284 for Edward I, and lay only about a day's ride away, but although Molières' plan shares the rectilinearity of Monpazier's it does not have the latter's orthogonality; see the plans in Lauret et al., *Bastides*, p. 205; Randolph, '*Bastides* of South-west France', pp. 291, 297. Similar differences are evident in the Edwardian new towns along the north coast of Wales, such as Flint and Rhuddlan.

80 See the plans in J. Passini, *El Camino de Santiago. Itinerarios y núcleos de poblacióu* (Madrid, 1993), p. 62; B. Gauthiez, *Atlas Morphologique des Villes de Normandie* (Lyon, 1999), p. 96.

81 Lilley, 'Taking Measures Across the Medieval Landscape', p. 85.

82 Lauret et al., *Bastides*, p. 81.

83 Conditions of topography and terrain are the usual reasons given to explain the distortions in regular layouts of medieval towns: see T. R. Slater, 'Ideal and Reality in English Episcopal Medieval Town Planning', *Transactions of the Institute of British Geographers*, new series, XII (1987), pp. 191–203. For arguments against this view see Lilley, 'Taking Measures Across the Medieval Landscape'.

84 Hager, *Zähringer New Towns*, np (section 'The Zähringer New Towns', para 7; 'Compositional elements', para 4).

85 Ibid. (section 'The Zähringer new towns', para 7; 'Compositional elements', para 5). The Berne plots measure sixty by one hundred feet (60:100 = 3:5). The plots at Breisach are evidently fifty by seventy-five feet (50:75 = 2:3).

86 For the extension see Hager, *Zähringer New Towns*, np. (caption for image 10, 'reconstruction by P. Hofer'). Hager does not note the repetition of the plot ratio in the layout of the extension.

87 Lilley, 'Urban Planning and the Design of New Towns', p. 17.

88 Lauret et al., *Bastides*, p. 66.

89 Friedman, *Florentine New Towns*, p. 73.

90 E. C. Fernie, 'The Romanesque Church of Bury St Edmunds Abbey', in *Bury St Edmunds*, ed. Gransden, p. 12. Fernie's analysis ignores the

irregularities present in the plan of Bury, see fig. 12, p. 13. See also Gauthiez, 'Planning of the Town of Bury St Edmunds', pp. 81–97.

91 They 'could well have been drawn using squares and root 2': Gauthiez, 'Planning of the Town of Bury St Edmunds', p. 89. See also his fig. 92.

92 Bucher, 'Architectural Design Methods', p. 43. Despite Bucher's claim I cannot see evidence for this in the Zähringer new towns. Bucher's interpretation of Grenade is contested by W. Boerefijn, 'Designing the Medieval New Town', *Urban Morphology*, IV (2000), pp. 49–62.

93 Friedman, *Florentine New Towns*, p. 132. Friedman took measurements on the ground at Grenade to confirm this idea of Bucher, and found 'that it was not the dimensions of the blocks that were controlled by geometry but the position of the survey lines by which the town was laid out'. For Villard see Bucher, *Architector*; and Victor, ed. and trans., *Practical Geometry in the High Middle Ages*. See also Lilley, 'Urban Planning and the Design of New Towns', pp. 239–40.

94 See J. Gimpel, *The Cathedral Builders*, trans. T. Waugh (London, 1983), p. 101; L. R. Shelby, 'The Geometrical Knowledge of the Mediaeval Master Masons', *Speculum*, XLVII (1972), pp. 395–421, at pp. 417–19.

95 Another plan that may be based on the same principle as Grenade's design is the new town of Castelfranco di Sotto, founded in the second half of the thirteenth century by the city of Lucca. A plan of it appears in Friedman, *Florentine New Towns*, p. 113. Friedman notes that 'three parallel files of blocks all had different proportions but were all rectangular' (p. 114), and yet the diagonal of the square blocks does seem to be about the same length as the rectangular blocks on either side of it.

96 Friedman, *Florentine New Towns*, pp. 8, 31.

97 Ibid., pp. 120–30.

98 Ibid., p. 75.

99 Eco, *Art and Beauty*, p. 40.

100 Aquinas, *Summa Theologia*, cited in Eco, *Art and Beauty*, p. 70.

101 Albert the Great, *De Pulchro et Bono*, trans. in Eco, *Aesthetics of Thomas Aquinas*, p. 112.

102 Robert Grosseteste, *Commentarium in Hexaemeron*, trans. in Eco, *Aesthetics of Thomas Aquinas*, p. 109.

103 Aquinas, *Summa Theologia*, cited in Eco, *Art and Beauty*, p. 71.

104 Eco, *Art and Beauty*, p. 40.

105 I. Bulmer Thomas, 'Euclid and Medieval Architecture', *Archaeological Journal*, CXXXVI (1979), pp. 136–50.

106 Vincent of Beauvais, *Speculum Doctrinale*, cited in Eco, *Art and Beauty*, p. 39.

107 Eco, *Art and Beauty*, pp. 33–5.

108 E. Grant, *The Foundations of Modern Science in the Middle Ages: Their Religious, Institutional, and Intellectual Contexts* (Cambridge, 1996), p. 46. See Stahl, *Roman Science*, pp. 134–69.

109 'If numbers are the key to harmony, they are likely to be the key to the entire universe': the Pythagorean basis of the 'doctrine of the haromy of the spheres', Stahl, *Roman Science*, p. 17. It is worth noting that the layout of Villeneuve-sur-Lot is made up of 3:4:5 triangles, see Lauret et al., *Bastides*, p. 81.

110 D. A. Callus, ed., 'Robert Grosseteste as Scholar', in *Robert Grosseteste: Scholar and Bishop*, ed. D. A. Callus (Oxford, 1955), pp. 1–69, at p. 26.

111 Taylor, trans., *Didascalicon of Hugh of St Victor*, pp. 141–2.

112 Isaiah 40:3–4, 42:16, 45:2, and Matthew 3:3, Mark 1:3, Luke 3:4–5; John 1:23, Hebrews 12:13.

113 D. Friedman, 'Palaces and the Street in Late Medieval and Renaissance Italy', in *Urban Landscapes: International Perspectives*, ed. J.W.R. Whitehand and P. J. Larkham (London, 1992), pp. 69–113.

114 See G. Lesser, *Gothic Cathedrals and Sacred Geometry* (London, 1957), I, p. 5, notes with regard to the 'geometric lucidity' to be found in the 'spatial structures' of Gothic cathedrals. More recently, see N. Hiscock, *The Wise Master Builder: Platonic Geometry in Plans of Medieval Abbeys and Cathedrals* (Aldershot, 2000); N. Hiscock, *The Symbol at Your Door: Number and Geometry in the Religious Architecture of the Greek and Latin Middle Ages* (Aldershot, 2008).

115 Cf. Mâle, *Religious Architecture.*

THREE: FOUNDING A CITY, FOUNDING A WORLD

1 Quoted in J. B. Friedman, 'The Architect's Compass in Creation Miniatures of the Later Middle Ages', *Traditio*, XXX (1974), pp. 419–29, p. 425, citing F. H. Colson and G. H. Whitaker, trans., *Philo*, vol. I. (Cambridge, MA, 1929), pp. 17–20.

2 Ibid.

3 See chapter One.

4 Genesis 4:17.

5 Genesis 2:18.

6 Hebrews 2:10.

7 See Introduction.

8 A. J. Gurevich, *Categories of Medieval Culture*, trans. G. L. Campbell (London, 1985), p. 163. For Aristotelian-influenced views of Christian kingship, see S. Ziller, C. Nederman and K. Forhan, ed. and trans. 'Thomas of Aquinas, *On Kingship*, *Summary of Theology* and *Commentary on Aristotle's Politics*', in *Medieval Political Theory – A Reader: The Quest for the Body Politic, 1100–1400*, ed. C. J. Nederman and K. Langdon Forhan (London, 1993), pp. 97–148, especially pp. 99–116. More generally see E. Kantorowicz, *The King's Two Bodies: A Study in Mediaeval Political Theology* (Princeton, NJ, 1957).

9 W. H. Stahl, *Roman Science: Origins, Development and Influence to the Later Middle Ages* (Madison, 1962).

10 A. Randolph, '*Bastides* of South-west France', *Art Bulletin*, LXXVII (1995), p. 306.

11 J. J. Sheridan, trans. *Alan of Lille: The Plaint of Nature* (Toronto, 1980), p. 120; see also P. E. Dutton, '*Illustre civitatis et populi exemplum*: Plato's *Timaeus* and the Transmission of Calcidius to the End of the Twelfth Century of a Tripartite Scheme of Society', *Mediaeval Studies*, XLV (1983), p. 114.

12 Alan of Lille, trans. in Dutton, '*Illustre civitatis et populi exemplum*', p. 114.

13 C. J. Nederman, ed. and trans., *John of Salisbury, Policraticus* (Cambridge, 1990), p. xvii; M. W. Beresford, *New Towns of the Middle Ages: Town Plantation in England, Wales and Gascony* (London, 1967), p. 331.

14 J. Dunbabin, 'The Reception and Interpretation of Aristotle's *Politics*', in *The Cambridge History of Later Medieval Philosophy: From the Rediscovery of Aristotle to the Disintegration of Scholasticism 1100–1600*, ed. N. Kretzman, A. Kenny and J. Pinborg (Cambridge, 1982), pp. 724–5.

15 J. M. Blythe, ed. and trans., *On the Government of Rulers, 'De Regimine Principum', Ptolemy of Lucca with Portions Attributed to Thomas Aquinas* (Philadelphia, PA, 1997), p. 97.
16 Ibid., pp. 96–7. Aquinas subsequently discusses some of these points in more detail, for example on 'the place for constructing a city' (p. 108 *et seq*), just before Ptolemy's text begins.
17 Cf. T. A. Sinclair, ed. and trans., *Aristotle, The Politics* (London, 1962), pp. 422–3.
18 See Friedman, 'Architect's Compass in Creation Miniatures'.
19 H. Bettenson, ed. and trans., *Augustine, The City of God* (London, 1984), p. 495. On the influence of Augustine and Boethius on Aquinas' aesthetics, see U. Eco, *The Aesthetics of Thomas Aquinas*, trans. H. Bredin (Cambridge, MA, 1988), pp. 71–81.
20 Blythe, ed. and trans. *On the Government of Rulers*, p. 96, cf. Book VII of Aristotle, *Politics*.
21 Ibid., pp. 39–40; J. H. Mundy, 'In Praise of Italy: The Italian Republics', *Speculum*, LXIV (1989), pp. 815–34, especially p. 833.
22 Blythe, ed. and trans. *On the Government of Rulers*, pp. 146, 220, cf. pp. 94–7.
23 On Aristotelian and Neoplatonic microcosmicism see Introduction. Ptolemy of Lucca draws further analogies between the body, the city-state and the universe in his continuation of Aquinas' *De Regimine*, see Blythe, ed. and trans. *On the Government of Rulers*, pp. 94–5, 104–8, 138, 179, 203, 221. See also H. Hamilton Bleakley, 'The Art of Ruling in Aquinas' *De Regimine Principum*', *History of Political Thought*, XX (1999), p. 577.
24 Blythe, ed. and trans. *On the Government of Rulers*, pp.10–11, 39–41.
25 D. Friedman, *Florentine New Towns: Urban Design in the Late Middle Ages* (Cambridge, MA, 1988), pp. 150–53. The nature of the autonomy of these committees changed from the later thirteenth to the later fourteenth centuries. Friedman notes (p. 151) that after *c.* 1315 came 'the formation of autonomous committees devoted to town-building', and that 'in 1332 six citizens received "all the powers of the priors and gonfaloniere" so that . . . the committee was relatively independent of the *Signoria*'.
26 See J. Dunbabin, 'Guido Vernani of Rimini's Commentary on Aristotle's "Politics"', *Traditio*, XLIV (1988), pp. 373–88, especially pp. 374, 376, 378.
27 Ziller et al., ed. and trans., 'Thomas of Aquinas, *On Kingship*', p. 98.
28 K. D. Lilley, 'Urban Landscapes and Their Design: Creating Town from Country in the Middle Ages', in *Town and Country in the Middle Ages*, ed. K. Giles and C. Dyer (Leeds, 2005), p. 232–8.
29 Ibid., p. 236.
30 Ibid., p. 232–3.
31 Randolph, '*Bastides* of South-west France', pp. 292–3.
32 Ibid., p. 300; Lilley, 'Urban landscapes and their Design', pp. 238.
33 See Beresford, *New Towns*.
34 R. Bartlett, *The Making of Europe: Conquest, Colonization and Cultural Change 950–1350* (London, 1993), pp. 24–59, 167–7.
35 K. D. Lilley, *Urban Life in the Middle Ages, 1000–1450* (London, 2002), pp. 106–37; Lilley, 'Urban Landscapes and their Design', p. 233.
36 M. Chibnall, *Anglo-Norman England 1066–1166* (Oxford, 1986), p. 148.
37 H. Quirin, 'The Colonial Town as Seen in the Documents of East German Settlement', *The Comparative History of Urban Origins in Non-Roman Europe*, ed. H. B. Clarke and A. Simms (Oxford, 1985), p. 509.

38 See examples cited in Introduction, above.
39 Friedman, 'Architect's Compass in Creation Miniatures'.
40 See chapter Four below. B. L. Ullman, 'Geometry in the Mediaeval Quadrivium', in *Studi de Bibliografia e di Storia in Onore di Tammaro de Marinis*, ed. T. de Marinis (Verona, 1964), vol. IV, pp. 263–85.
41 More will be said on this later. For some English examples of named architects involved in construction work in the thirteenth and fourteenth centuries see J. Harvey, *English Mediaeval Architects: A Biographical Dictionary Down to 1550* (Gloucester, 1984); R. A. Brown, H. M. Colvin and A. J. Taylor, *History of the King's Works*, I and II*: The Middle Ages* (London, 1963). For some Italian examples see Friedman, *Florentine New Towns*, pp. 149–66.
42 For examples, see J. K. Hyde, 'Medieval Descriptions of Cities', *Bulletin of the John Rylands Library*, XLVIII (1966), pp. 308–40, at p. 308.
43 J. Gimpel, *The Cathedral Builders*, trans. T. Waugh (London, 1983), pp. 89–121, p. 121.
44 See G. Binding, *Medieval Building Techniques*, trans. A. Cameron (Stroud, 2004), see examples numbered 3a, 34a, 47, 48, 89, 124, 166, 168, 200–201, 209–10, 254, 260, 267, 276a, 278, 281, 284, 286, 301, 328, 409, 434, 467, 498, 507f, 538a, 556, 558, 602, 623, 627, 648b. Some illustrations show architects working at drawing boards, for example 248 and 500.
45 Gimpel, *Cathedral Builders*, pp. 117–19; R. Branner, 'Villard de Honnecourt, Reims and the Origin of Gothic Architectural Drawing', *Gazette des Beaux-Arts*, LXI (1963), pp. 129–46; D. Friedman, 'Palaces and the Street in Late Medieval and Renaissance Italy', in *Urban Landscapes: International Perspectives*, ed. J.W.R. Whitehand and P. J. Larkham (London, 1992), pp. 93–6.
46 F. Bucher, *Architector: The Lodgebooks and Sketchbooks of Medieval Architects*, vol. I (New York, 1979), p. 112. See also S. K. Victor, ed. and trans., *Practical Geometry in the High Middle Ages: 'Artis Cuiuslibet Consummatio' and the 'Pratike de Geometrie'* (Philadelphia, PA, 1979), pp. 68–73; Branner, 'Villard de Honnecourt', pp. 140–41.
47 R. Steele, ed., *Communia mathematica fratris Rogeri [Roger Bacon]* (Oxford, 1940), p. 42. Victor, ed. and trans., *Practical Geometry in the High Middle Ages*, p. 57.
48 Victor, ed. and trans., *Practical Geometry in the High Middle Ages*, pp. 23–4, 213, 215.
49 Drawings of cities were used in maps, for example, including the so-called Gough Map of Great Britain, contemporary with the time of Bacon. In the same list of geometry's uses, and following on directly from saying it is used to draw cities, Bacon comments that geometry is also used in 'how to make canals etc' and 'how to construct machines etc': Steele, ed., *Communia mathematica fratris Rogeri*, p. 42.
50 P.D.A. Harvey, 'Local and Regional Cartography in Medieval Europe', in *The History of Cartography*, vol. I: *Cartography in Prehistoric, Ancient and Medieval Europe and the Mediterranean*, ed. J. B. Harley and D. Woodward (Chicago, IL, 1987), p. 492, figure 20.27. For examples of plans drawn of existing plots see M. G. Snape, 'Durham 1439 x circa 1442', in *Local Maps and Plans from Medieval England*, ed. R. A. Skelton and P.D.A. Harvey (Oxford, 1986), p. 189–94.
51 U. T. Holmes, *Daily Living in the Twelfth Century: Based on the Observations of Alexander Neckam in London and Paris* (London, 1952), p. 275. In London in the early 1400s an attempt was made to divide up Moorfields

(an area outside the city walls) into gardens 'to be let out to the profit of the city chamber', a decision which not only involved the mayor and alderman but also 'an immense congregation of very reputable men of the city', and done 'in accordance with a plan set out on a sheet of parchment and shown to those present at the meeting'. The plan has not survived. See C. M. Barron, *London in the Later Middle Ages: Government and People 1200–1500* (Cambridge, 2004), p. 246.

52 L. Shopkow, ed. and trans., *Lambert of Ardres, The History of the Counts of Guines and Lords of Ardres* (Philadelphia, PA, 2001), p. 190. For an alternative translation see C. Frayling, *Strange Landscape: A Journey through the Middle Ages* (London, 1995), p. 64.

53 Thomas Aquinas, *Summa Theologica*, cited in Eco, *Aesthetics of Thomas Aquinas*, p. 101. Aquinas is discussing what is 'true'.

54 Friedman, *Florentine New Towns*, pp. 153, 155.

55 Ibid., pp. 155–7: '*Magistrum et cepudmagistrum et proveditorem constructionem dicte terre Vicchi*', in the case of Cambiuzzi's work at Vicchio (p. 155).

56 Ibid., pp. 156–9, 162–3.

57 Ibid., p. 158.

58 E. C. Fernie, 'The Romanesque Church of Bury St Edmunds Abbey', in *Bury St Edmunds*, ed. Gransden; B. Gauthiez, 'The Planning of the Town of Bury St Edmunds: A Probable Norman Origin', in *Bury St Edmunds*, ed. Gransden.

59 See above. The use of rotating squares (*ad quadratum*) in the 'hidden' geometry of the town's design suggests someone with knowledge of architectural theory had been responsible for designing Grenade.

60 Besides the measurers and architects, a third group of individuals who were evidently working in towns and cities on building and planning work were engineers, who also would have expertise in geometry and familiarity with *practica geometriae* texts. See Lilley, 'Urban Landscapes and their Design', pp. 235–6.

61 See J. Schulz, 'Jacapo de' Barbari's View of Venice: Map Making, City Views and Moralized Geography before the Year 1500', *Art Bulletin*, LX (1978), p. 432, n. 20. On Siena see W. M. Bowsky, *A Medieval Italian Commune: Siena under the Nine, 1287–1355* (Berkeley, CA, 1981), pp. 260–98; on Florence see Friedman, 'Palaces and the Street', p. 207.

62 A. de Smet, 'De l'utilité de recueillir les mentions d'arpenteurs cités dans les documents d'archives du moyen âge', *Annales du Congres-Archéologique et Historique de Tournai*, III (1949), pp. 782–95.

63 Ibid., p. 785.

64 Ibid., pp. 783, 787.

65 J. C. Tingey, ed., *The Records of the City of Norwich*, vol. II (Norwich, 1910), pp. 32–7, 52.

66 Frayling, *Strange Landscape*, p. 64, suggests this phrase translates as 'so learned in geometrical work'.

67 J. P. Trabut-Cussac, 'Date, fondation et identification de la *bastide* de Baa', *Revue Historique de Bordeaux*, X (1961), pp. 133–44, at p. 142. In the context of work at the Florentine town of San Casciano, Giovanni di Lapo Ghini is referred to as '*magister ordinatoris*', and Friedman, *Florentine New Towns*, p. 156, notes that '"*ordinare*" is the act of physical planning'.

68 Trabut-Cussac, 'La *bastide* de Baa', pp. 142–3. Victor, ed. and trans., *Practical Geometry in the High Middle Ages*, p. 63, seems to misinterpret Richard de Escham's role as he suggests Escham 'was paid for laying out the streets', whereas Master Gérard would seem to have been the

one doing the laying out.

69 As would appear to be the case with the often cited example of New Winchelsea (Sussex), where its refounding on a new site by the order of Edward I in the 1280s is usually seen (e.g., by Beresford, *New Towns*, pp. 14–28; Friedman, *Florentine New Towns*, pp. 149–50) to have been the work of those the king appointed, that is Stephen de Pencestre, Gregory de Rokesle and Henry le Waleys, but they would surely have delegated the task to someone qualified to do it, such as a measurer. See Lilley, 'Urban Landscapes and their Design', pp. 232–6.

70 Leonardo Fibonacci, *Practica Geometriae*, in Victor, ed. and trans. *Practical Geometry in the High Middle Ages*, p. 47. See also Friedman, *Florentine New Towns*, p. 125.

71 Victor, ed. and trans. *Practical Geometry in the High Middle Ages*, pp. 47–9.

72 Friedman, *Florentine New Towns*, p. 129.

73 H.L.L. Busard, ed., 'The *Practica Geometriae* of Dominicus de Clavasio', *Archive for the History of Exact Sciences*, II (1965), pp. 520–75, at pp. 559–60. See Victor, ed. and trans. *Practical Geometry in the High Middle Ages*, p. 52.

74 Cited in T. Zagrodski, 'Les plans de villes créées au moyen âge en Pologne et la tradition des règles gromaticales', *Revue des Archéologiques et Historiens d'Art de Louvain*, IX (1976), p. 118.

75 Randolph, '*Bastides* of South-west France', p. 300, writes: 'at the foundation ceremony, it [the *palum*] was raised at the centre of the market place as the embodiment of the charter, the contents of which were made known to all through the announcements of a town crier'. He notes this occurred with the foundation of Pamplonne in 1292: 'Eustache de Beaumarchais planted the *pau* [pole] in the place which was once called Thuria and he built a new *bastide* to which he gave the name Pampilona', citing A. Curie-Seimbres, *Essai sur les Villes Fondées dans le Sud-Ouest de la France aux XIIIe et XIVe Siècles sous le Nom Générique de Bastides* (Toulouse, 1880), p. 161. This, however, would seem to indicate the *palum* was planted at the start of the foundation process, to mark the inauguration of the process rather than its ending. It is worth noting here too that it was Eustache de Beaumarchais who founded Grenade-sur-Garonne at this same time, with its proportioned geometrical design.

76 The grant of the charter was the end of the town's beginning, making it dubious to use 'foundation' charters to determine the date at which a town originated. Charters often were granted well after the town had been physically made. See Lilley, 'Urban Landscapes and their Design', pp. 237–8.

77 Beresford, *New Towns*, pp. 290–315, discusses the relative economic successes of the new towns of England, Wales and Gascony.

FOUR: MEASURES OF MEANING

1 F. A. Homann, trans., *Practical Geometry [Practica Geometriae] Attributed to Hugh of St Victor* (Milwaukee, WI, 1991); R. Baron, 'Hughes de Saint-Victor, auteur d'une *Practica Geometriae*', *Mediaeval Studies*, XVII (1955), pp. 107–16, at p. 116 suggests a date of authorship of around 1125–30. See also P. Tannery, 'La géométrie au XIe siècle', *Revue Générale Internationale Scientifique, Litteraire et Artistique*, XV (1897), pp. 343–57; and P. Tannery, 'Géométrie', *Revue de Synthèse Historique*, II(1901), pp. 283–99; S. K. Victor, ed. and trans., *Practical Geometry in the High Middle Ages:*

'Artis Cuiuslibet Consummatio' and the 'Pratike de Geometrie' (Philadelphia, PA, 1979), pp. 3–7, 17–19.

2 Homann, trans. *Practical Geometry*, p. 34. Hugh's division is the same as that set out by Dominicus Gundissalinus, *De divisione philosphiae*, as noted by L. R. Shelby, 'The Geometrical Knowledge of the Mediaeval Master Masons', *Speculum*, CXLVII (1972), pp. 402–3. Shelby suggests Gundissalinus was influenced by Hugh's *Practica geometriae*, as well as Al-Fārābī's *De scientiis*: Shelby quotes Gundissalinus as follows (p. 403): 'there are three species of practical geometry: altimetry, planimetry, cosmimetry. That science by which one considers lines, surfaces and bodies in height is called altimetry, that is to say, the science of measuring altitudes; in planes it is called planimetry, that is, the science of measuring any plane surface; in depth, it is called cosmimetry or the science of measuring solids'. See also Victor, ed. and trans., *Practical Geometry in the High Middle Ages*, pp. 7–12. On Al-Fārābī, see D. Salmon, 'The Mediæval Latin Translations of Alfarabi's works', *New Scholasticism*, XIII (1939), pp. 245–61. Homann, trans. *Practical Geometry*, p. 34, suggests this tripartite division of geometry traced back to Nipsus and the Roman agrimensores.

3 Homann, trans., *Practical Geometry*, pp. 35–6.

4 Ibid., pp. 49 (on 'less skillful surveyors'), 59, 70 (on Macrobius). Hugh also refers to the use of models, which also appears in W. H. Stahl, trans., *Macrobius' Commentary on the Dream of Scipio* (New York, 1952), pp. 175, 202.

5 Victor, ed. and trans., *Practical Geometry in the High Middle Ages*, p. 25; see also p. 19 for comparison with High of St Victor's 'cosmimetry' and the *Artis* term *'crassimetria'*.

6 Ibid., p. x.

7 Ibid., pp. 335, 115.

8 Ibid., pp. 211–19.

9 On the *Sphere* and its commentators see L. Thorndike, ed. and trans., *The 'Sphere' of Sacrobosco and its Commentators* (Chicago, IL, 1949); on the 'Treatise of the Quadrant', Victor, ed. and trans., *Practical Geometry in the High Middle Ages*, p. 21.

10 Thorndike, ed. and trans., *Sphere of Sacrobosco*, pp. 1, 118–19.

11 Homann, trans., *Practical Geometry*, p. 24 remarks that although the readers of Hugh of St Victor's *practica geometriae* 'are not identified, the use of Latin suggests schoolmen, while the content suggests surveyors and technicians'. Victor, ed. and trans., *Practical Geometry in the High Middle Ages*, pp. 54–5, likewise concurs that these texts were written to inform and educate. Since the authors themselves differentiated between types of reader, as with the texts by Fibonacci and Clavasio, this was surely the case (see above). Cf. Shelby, 'Geometrical Knowledge'.

12 On the geometrical content of the Quadrivium see B. L. Ullman, 'Geometry in the Mediaeval Quadrivium', in *Studi de Bibliografia e di Storia in Onore di Tammaro de Marinis*, ed. T. de Marinis (Verona, 1964), IV, pp. 263–85; W. H. Stahl, *Roman Science: Origins, Development and Influence to the Later Middle Ages* (Madison, 1962), also more generally, D. L. Wagner, ed., *The Seven Liberal Arts in the Middle Ages* (Bloomington, 1983).

13 L. R. Shelby, 'The Education of Medieval English Master Masons', *Mediaeval Studies*, XXXII (1970), pp. 1–26; the widespread use of texts is revealed by the number of manuscript copies that are still in existence.

The 'Treatise on the Quadrant' survives in seventy-five manuscripts for example: Victor, ed. and trans., *Practical Geometry in the High Middle Ages*, p. 21. Shelby notes the importance of oral transmission of geometrical knowledge between masters and patrons in the Middle Ages, see Shelby, 'Geometrical Knowledge', p. 398.

14 Shelby, 'Geometrical Knowledge', p. 401.

15 Wagner, ed., *Seven Liberal Arts*; Ullman, 'Geometry in the Medieval Quadrivium', p. 263.

16 C. Burnett, ed., *Adelard of Bath: An English Scientist and Arabist of the Early Twelfth Century* (London, 1987); H.L.L. Busard, ed., *The First Translation of Euclid's 'Elements', Commonly Ascribed to Adelard of Bath* (Toronto, 1983).

17 On geometry and Boethius see Stahl, *Roman Science*, pp. 200–201.

18 Watts, ed. and trans. *Boethius Consolation*, p. xvii, on Boethius' *Arithmetica*.

19 Stahl, *Roman Science*, p. 201, citing T. L. Heath, *History of Greek Mathematics*, vol. I (Oxford, 1921), p. 359.

20 Stahl, *Roman Science*, p. 201. Stahl gives the date range as tenth to fourteenth century for these, while Ullman, 'Geometry in the Mediaeval Quadrivium', pp. 267, 270, gives ninth to thirteenth centuries as the range.

21 Ullman, 'Geometry in the Mediaeval Quadrivium', pp. 267–78, 270. He cites in particular the following 'excerpt manuscripts': Naples (V.A. 13) ('N') of the ninth or tenth century; Trinity College, Cambridge (939) ('T') of the tenth century; Paris Bibliothèque Nationale (BN. 13955) ('p2') of the tenth century; St Gall (830) (g) of the eleventh century; Berne (87) ('ba') of 1004; and British Museum (BM Add. 47679) of the twelfth century. Thulin describes these as the XYZ manuscripts in his study of the *Corpus Agrimensores Romanorum*: C. Thulin, ed., *Corpus Agrimensorum Romanorum*, vol. I, fasc. I, *Opuscula agrimensorum veterum* (Leipzig, 1913), pp. iii–iv.

22 Ullman, 'Geometry in the Mediaeval Quadrivium', pp. 266, 269 argues that the excerpt manuscripts were for university audiences and 'not to produce surveyors'. See also E. A. Zaitsev, 'The Meaning of Early Medieval Geometry, from Euclid and Surveyors' Manuals to Christian Philosophy', *Isis*, XC (1999), pp. 530–35.

23 The gromatic texts of the twelfth and thirteenth centuries often cited Euclid or used Euclidian propositions to prove a particular principle in measurement or surveying work. One of the most 'Euclidian' of these, Leonardo Fibonacci's *Practical Geometry*, certainly did draw upon Euclidian theorems and propositions, but some authors, as Victor has suggested, simply cited Euclid as a source because they associated his name with geometry and not necessarily because they themselves had consulted the *Elements*. So potentially at least, the *Elements* influenced those who were using these texts in their practical work. See Victor, ed. and trans., *Practical Geometry in the High Middle Ages*, pp. 46, 50–51; also Shelby, 'Geometrical Knowledge', p. 404 on Leonardo Fibonacci. The *Artis* text shows explicit use of Euclid's theorems, see Victor, ed. and trans., *Practical Geometry in the High Middle Ages*, pp. 83, 115, 117. It has been argued by I. Bulmer Thomas in 'Euclid and Medieval Architecture', *Archaeological Journal*, CXXXVI (1979) that the wider knowledge of Euclid's *Elements* after Adelard's translation was a factor in the straightening up of the ground-plans of churches in the later

twelfth and thirteenth centuries. Exactly how this application of Euclidian geometry might have occurred would require a close examination of the *practica geometria* texts written after *c.* 1150, though at least one critic has argued against seeing the translation of the *Elements* as important in the state of practical geometrical knowledge: Shelby, 'Geometrical Knowledge', p. 420. Whether or not the *Elements* itself had an influence on the material shapes of buildings or indeed urban landscapes, what is clear all the same is that theoretical geometry, derived from Euclid, was associated with the wider cosmos whether it came via Boethius or the later *practica geometriae*, and gave those who used geometry for practical ends an awareness of its cosmic symbolism. See Zaitsev, 'Meaning of Early Medieval Geometry', pp. 523–30.

24 Theoretical geometry derived more from Greek geometry, especially Euclid, while practical geometry derived more from Roman surveying work. See Ullman, "Geometry in the Mediaeval Quadrivium", p. 264, on theoretical and practical geometries; also R. Baron, 'Sur l'introduction en Occident des termes 'geometria theorica et practica", *Revue d'histoire des sciences et de leurs applications*, VIII (1955), p. 298; and Shelby, 'Geometrical Knowledge', p. 401.

25 Homann, trans., *Practical Geometry*, p. 33.

26 Shelby, 'Geometrical Knowledge', p. 403, citing L. Baur, ed., *Dominicus Gundissalinus, 'De divisione philosophiae'* (Münster, 1903), pp. 107–110. On Dominicus and geometry see Victor, ed. and trans., *Practical Geometry in the High Middle Ages*, p. 12.

27 Cited in A. Drew, 'The *De Eodem et Diverso*', in *Adelard of Bath: An English Scientist and Arabist of the Early Twelfth Century*, ed. C. Burnett (London, 1987), pp. 17–24, at p. 23; see Victor, ed. and trans., *Practical Geometry in the High Middle Ages*, p. 13.

28 See Homann, trans., *Practical Geometry.*

29 Gerbert's geometriae see Tannery, 'Géométrie au XI siècle', pp. 352–4, 56–7; N. Bubnov, ed., *Gerberti Opera mathematica (972–1003)* (Berolini, 1899).

30 Shelby, 'Geometrical Knowledge', pp. 400–1. Ullman, 'Geometry in the Mediaeval Quadrivium'.

31 B. Campbell, ed. and trans., *The Writings of the Roman Land Surveyors: Introduction, Text, Translation and Commentary* (London, 2000).

32 O.A.W. Dilke, 'Illustrations from Roman Surveyors' Manuals', *Imago Mundi*, XXI (1967), pp. 9–29, at pp. 25–7, citing Hyginus Gromaticus. For Hyginus, see B. Campbell, 'Shaping the Rural Environment: Surveyors in Ancient Rome', *Journal of Roman Studies*, LXXXVI (1996), pp. 74–99, at pp. 77–8.

33 J. N. Carder, *Art Historical Problems of a Roman Land Surveying Manuscript: The Codex Arcerianus A, Wolfenbüttel* (New York, 1978), pp. 1–2.

34 For example from Isidore and Boethius. Carder, *Art Historical Problems*, pp. 2–7. The earliest manuscripts of the Corpus are the Arcerianus A and B recensions.

35 For the Palatinus and so-called codex mixti manuscripts, see Carder, *Art Historical Problems*, pp. 4, 8–9, 16, 27; Campbell, ed. and trans., *Writings of the Roman Land Surveyors*, p.xxii; Ullman, 'Geometry in the Mediaeval Quadrivium', p. 275; Shelby, 'Geometrical Knowledge', p. 400.

36 For an overview see Campbell, ed. and trans., *Writings of the Roman Land Surveyors.*

37 See Zaitsev, 'Meaning of Early Medieval Geometry'.

38 Homann, trans., *Practical Geometry*, pp. 4, 7, 15–17.
39 Ullman, 'Geometry in the Mediaeval Quadrivium', pp. 266–71. He notes that the *ars geometriae* 'contains much material intended for surveyors', p. 270.
40 Ibid., pp. 266–7, 269.
41 Homann discusses this in relation to the 'plan' of St Gall, see Homann, trans., *Practical Geometry*, pp. 12–15. Carder, *Art Historical Problems*, p. 2, also discusses the pedagogical purpose of the illustrations that were part of the *Corpus* texts. Illustrations were a means of visually communicating the surveying techniques though as Carder shows, not all the manuscripts had illustrations and in some cases they seem to be simply there as embellishments of the text rather than instructive diagrams: Carder, *Art Historical Problems*, pp. 108, 167. Illustrations continued to be used in the later *practica geometriae* texts: see Bucher, *Architector*; Victor, ed. and trans., *Practical Geometry in the High Middle Ages*.
42 See Zaitsev, 'Meaning of Early Medieval Geometry', pp. 535–46.
43 Campbell, ed. and trans., *Writings of the Roman Land Surveyors*; Campbell, 'Shaping the Rural Environment'.
44 Campbell, 'Shaping the Rural Environment', p. 82; J. Rykwert, *The Idea of a Town: The Anthropology of Urban Form in Rome, Italy and the Ancient World* (London, 1988), p. 50.
45 Campbell, 'Shaping the Rural Environment', p. 84.
46 Ibid., plate 1.
47 Homann, trans., *Practical Geometry*, pp. 75–6: Hugh's planimetry exercise was based on agrimensorial practice.
48 Rykwert, *Idea of a Town*, p. 62. In particular he points to the importance of boundaries in Roman law and the role of surveyors both in marking them out and recording them, as well as the act of founding a town through ploughing a furrow in cosmic imitation, as Roman and Neoplatonic sources such as Varro, Plutarch and Macrobius reveal. Rykwert, *Idea of a Town*, pp. 45–7, 62, 65.
49 Hyginus 2 in Campbell, ed. and trans., *Writings of the Roman Land Surveyors*, p. 135.
50 This involved tracing out a diagram on the ground, the *templum*, which was itself an image of the heavens, divided into quarters and orientated cardinally, and began with the augur who performed the foundation rite and the surveyors who traced out the *limites*. Rykwert, *Idea of a Town*, p. 46.
51 Frontinus in Campbell, ed. and trans., *Writings of the Roman Land Surveyors*, p. 9.
52 Ibid., p. 13; Hyginus explains the *groma* used to establish the *limites*: see Hyginus 2 in Campbell, ed. and trans. *Writings of the Roman Land Surveyors*, p. 137. The *groma* fitted onto the *ferramentum*, see M.J.T. Lewis, *Surveying Instruments of Greece and Rome* (Cambridge, 2001), p. 126.
53 Hyginus Gromaticus; see Lewis, *Surveying Instruments*, pp. 22–3, 125; Dilke, 'Illustrations from Roman Surveyors' Manuals', p. 9, where he notes 'the name of the Roman cross-staff, groma, is evidently an Etruscan adaptation of Greek (*gnoma*), a by-form of *gnomon*'.
54 See Carder, *Art Historical Problems*, pp. 62–4, 90–92, illus 52 (Ms. A, f. 42r), 56 (Ms. A, f. 42v), 107 (Ms. P, f. 89v), 111 (Ms. P, f. 90r) and 113 (Ms. P, f. 90r).
55 T. Zagrodski, 'Les plans de villes créées au moyen âge en Pologne et la tradition des règles gromaticales', *Revue des Archéologiques et Historiens*

d'Art de Louvain, IX (1976), pp. 108–9; Randolph, '*Bastides* of South-west France', pp. 299–300.

56 W. Horn and E. Born, ed., *The Plan of St. Gall: a Study of the Architecture and Economy of Life in a Paradigmatic Carolingian Monastery* (Berkeley, 1979), 3 vol..

57 Dilke, 'Illustrations from Roman Surveyors' Manuals', p. 11. The Arcerianus manuscript was at Bobbio in 1493, and 'it seems that St Columban (543–615), first abbot of Bobbio, who originally came from Ireland, acquired the library of Cassiodorus (*c.*487–583), which at the statesman's death was in the monastery of Vivarium. Among the letters of Cassiodorus is one . . . in which he shows knowledge of surveying . . . We may think it probable, therefore, that he possessed a collection of Latin surveying manuals, which are likely to have found their way to Bobbio', writes Dilke, 'Illustrations from Roman Surveyors' Manuals', pp. 11–12. See also Ullman, 'Geometry in the Mediaeval Quadrivium'.

58 Victor, ed. and trans., *Practical Geometry in the High Middle Ages*, p. 83. It is also worth noting the parallels between Hugh's diagrams of the quartered circular world and those in the Corpus, for example in Hyginus and Frontinus. See Homann, trans. *Practical Geometry*, pp. 36–7.

59 Cited in E. R. Kiely, *Surveying Instruments: Their History and Classroom Use* (New York, 1947), p. 63.

60 See Carder, *Art Historical Problems*, p. 16, citing Bubnov, ed., *Gerberti Opera mathematica*, pp. 439, 475.

61 See Lewis, *Surveying Instruments*, pp. 22–23. For example, the gnomon's use for orientation is described by Pliny, *Natural History* and Vitruvius, *On Architecture*: see Rykwert, *Idea of a Town*, pp. 48–49. Obrist, 'Wind Diagrams and Medieval Cosmology', p. 43, notes a manuscript illustration of *c.*800 of Vitruvius' wind diagram. Hugh of St Victor refers to both Pliny and Vitruvius in his *Didascalicon* showing he was familiar with their works and hence their ideas on the uses of the gnomon. Cf. J. Taylor, trans., *The Didascalicon of Hugh of St Victor: A Medieval Guide to the Arts* (New York, 1961), pp. 83, 84.

62 Dilke, 'Illustrations from Roman Surveyors' Manuals', p. 17.

63 For a 'shadow stick', see Lewis, *Surveying Instruments*, pp. 22–3; Kiely, *Surveying Instruments*, pp. 61–2.

64 See Rykwert, *Idea of a Town*, pp. 48–9.

65 For example, Zagrodsky, 'Plans de villes', pp. 108–9; A. Randolph, '*Bastides* of South-west France', *Art Bulletin*, LXXVII (1995), p. 300.

66 Homann, trans., *Practical Geometry*, p. 59. Hugh also uses the gnomon 'to calculate the size of the solar sphere and its diameter' (p. 60), a function which Homann further discusses, writing: Hugh's '*Practical Geometry* text of the gnomon shadow method attributed to Eratosthenes is found almost verbatim in *Geometria incerti auctoris* . . . Bubnov cites Martianus Capella's *De nuptiis philogiae et mercurii* as a reference for this, but remarks that *incerti auctor* seems to have based his text not on Martianus but rather on one of the *Gromatici Veteres* (the Roman surveyors), whose manuscript he had at hand' (p. 82). See Bubnov, ed., *Gerberti Opera mathematica*, pp. 362–3. Also Stahl, *Roman Science*, p. 239.

67 Victor, ed. and trans., *Practical Geometry in the High Middle Ages*, pp. 323–5: the *Artis* text does not actually mention the gnomon by name and refers to it simply as 'an instrument for having the shadow of the altitude', but the method described (and illustrated) is that known as the Indian circles method referred to by Gerbert 'in setting out of the *cardo* and *decimanus*'

(Kiely, *Surveying Instruments*, p. 62). Victor shows the close connections between the *Artis* text and Gerbert's geometry (p. 83).

68 H.L.L. Busard, ed., 'The *Practica Geometriae* of Dominicus de Clavasio', *Archive for the History of Exact Sciences*, II (1965), p. 534. Busard (p. 522) notes Dominicus' debt to Gerbert in the description of the gnomon. Dominicus also describes the quadrant and astrolabe (pp. 530–31).

69 See Ullman, 'Geometry in the Mediaeval Quadrivium', p. 283, who says 'Corbie was the gromatic and geometric capital of the medieval world' and gives his reasons why. As well as Hugh's, other works on geometry were being composed in northern France, in the eleventh century for example, again with a Gerbertian influence present – see Tannery, 'Géométrie au XI siècle', pp. 347–54.

70 Kiely, *Surveying Instruments*, p. 61: 'we may safely assume that no form of surveyor's cross was in common use during this period'.

71 H. Mendthal, ed., *'Geometria culmensis', ein agronomischer Traktat aus der Zeit des Hochmeisters Conrad von Jungingen (1393–1407)* (Leipzig, 1886).

72 Zagrodski, 'Plans de villes', p. 108; Kiely, *Surveying Instruments*, p. 61, citing Mendthal, ed., *Geometria culmensis*, p. 31: 'Duch sal man allewege czulegen das rechte winkelmos adir crucze, wornoch man misset, wenne alle mose geen uo dem rechten winkel'.

73 Zagrodski, 'Plans de villes', pp. 109–18. With respect to the layouts of medieval towns in England, a view shared by E. Guidoni, *La ville Européenne: Formation et signification du quatrième au onzième siècle* (Brussels, 1981), p. 161.

74 R. T. Gunther, *Early Science in Oxford*, vol. I (Oxford, 1923), p. 338, where he explains how the quadrant was used to create perpendicular lines.

75 Kiely, *Surveying Instruments*, pp. 50–53, 64–75.

76 Ibid., pp. 50, 53, 74–5, 77–83; Gunther, *Early Science*, pp. 337–39.

77 Victor, ed. and trans., *Practical Geometry in the High Middle Ages*, pp. 115, 117, 121, 189, 317; Homann, trans., *Practical Geometry*, pp. 40, 42–3, 45–7, 49, 50–51, 54–5, 57, 59–60; Kiely, *Surveying Instruments*, p. 79. Victor notes (p. 121, note 14) that 'strictly speaking, the astrolabe and quadrant are used differently in finding distances. In a quadrant one sights along the edge of the quadrant as it pivots around its corner; the reading is taken where a plumb-line is attached, crosses the edge of the shadow quadrant. In an astrolabe, the shadow quadrant remains fixed while the alidad moves. Once the number of digits has been read, the computation is the same'. See also D. Friedman, *Florentine New Towns: Urban Design in the Late Middle Ages* (Cambridge, MA, 1988), pp. 138–42.

78 P. Tannery, 'Le Traité du quadrant de maître Robert Anglès', *Notices et extraits des manuscrits de la Bibliothèque Nationale*, XXXV (1897), pp. 561–640, cited in Thorndike, ed. and trans., *Sphere of Sacrobosco*, p. 63. A date of 1276 is given to the *Quadrans vetus.*

79 Gunther, *Early Science*, pp. 336–8, referring to Bib. Sloan. BM. 213, xiv, f.120. See J. Halliwell, ed., *Rara Mathematica* (London, 1839), pp. 56–71.

80 Thorndike, ed. and trans., *Sphere of Sacrobosco*, p. 4. For Robertus Anglicanus, 'Commentary on the Sphere', see Thorndike, ed. and trans., *Sphere of Sacrobosco*, pp. 211–12 (Robert mentions the astrolabe, too, as does Sacrobosco though curiously he does not likewise mention the quadrant, see p. 122). On the date of Robert's 'Commentary', see p. 28.

81 Stahl, *Roman Science*, pp. 238–9, who notes Gerbert's time spent learning in Barcelona, his teaching at Reims in 972 and appointment as Pope Silvester II in 999. Gerbert sent a letter to Bishop Lupitus (Lopez) of

Barcelona in 984 'requesting him to send a copy of *Liber de astrologia* that Lopez had translated from Arabic' (p. 239). A *Liber de astrolabio* is attributed to Gerbert and in the eleventh century further treatises on the use of the astrolabe appeared in France.

82 Homann, trans., *Practical Geometry*, p. 25; Kiely, *Surveying Instruments*, pp. 74–5.

83 Victor, ed. and trans., *Practical Geometry in the High Middle Ages*, pp. 114–21, where the *Artis* author examines how 'to measure the quantity of a straight line' using an astrolabe. He also proposes the same task can be undertaken using sticks and with the quadrant. The source for this is one of the 'inauthenticated' Gerbertian texts, *Geometria incerti auctoris*, see Victor, ed. and trans., *Practical Geometry in the High Middle Ages*, p. 115, n. 9.

84 This difference in approach is suggested by Dominicus Clavasio and his differentiation between the *mensor geometrie* and the *mensor laicus*, in his *Practica Geometriae*; that 'the difference between them lies in 'that what the lay measurer knows how to measure by going and dashing around the sides of a field with his rods and cords, the geometric measurer will know standing still by mental reflection or by drawing lines': Busard, '*Practica geometriae* of Dominicus Clavasio', pp. 559–60; Victor, ed. and trans., *Practical Geometry in the High Middle Ages*, p. 52.

85 An absence noted by Shelby, 'Geometrical Knowledge', p. 407.

86 See Kiely, *Surveying Instruments*, pp. 64–75, 77–83.

87 Roger Bacon, *Opus majus*, cited in Victor, ed. and trans., *Practical Geometry in the High Middle Ages*, pp. 32–7. Victor discusses the geometrical exegeses of Hugh and Richard, citing in particular Hugh's *De arca Noe morali* and Richard's *In visionem Ezechielis*. See also D. Woodward, 'Medieval *mappaemundi*', in *The History of Cartography*, vol. I: *Cartography in Prehistoric, Ancient and Medieval Europe and the Mediterranean*, ed. J. B. Harley and D. Woodward (Chicago, IL, 1987), p. 335, for Hugh's account of the mystical Noah's ark, where he notes the microcosmism of the passage as well as its 'account of the making of a *mappamundi*'.

88 Victor, ed. and trans., *Practical Geometry in the High Middle Ages*, p. 33.

89 Taylor, trans., *Didascalicon of Hugh of St Victor*, p. 140.

90 Ibid., pp.140–42.

91 See Zaitsev, 'Meaning of Early Medieval Geometry', pp. 535–9; Friedman, 'Architect's Compass in Creation Miniatures', pp. 419–21.

92 See L. R. Shelby, 'Medieval Masons' Tools II. Compass and Square', *Technology and Culture*, VI (1965), pp. 236–48.

93 J. B. Friedman, 'The Architect's Compass in Creation Miniatures of the Later Middle Ages', *Traditio*, XXX (1974), p. 426.

94 Proverbs 8:27.

95 Rashi uses the Old French 'kompas': Friedman, 'Architect's Compass in Creation Miniatures', p. 426. Cf. Isaiah 44:13 describing a carpenter at work on 'the figure of a man'.

96 B. Stock, *Myth and Science in the Twelfth Century: A Study of Bernard Silvester* (Princeton, NJ, 1972), p. 253, citing É. Jeauneau, ed., *Guillaume de Conches, 'Glosae super Platonem'* (Paris, 1975), pp. 266–7. William comments that 'God is like a good craftsman'; also Stock, *Myth and Science*, pp. 264–5 on Daniel of Morley, *Liber de naturis inferiorum et superiorum*.

97 In Christian art of the Middle Ages, the compass was used to symbolize geometry. For example, see E. Mâle, *Religious Art in France: The Thirteenth Century* (Princeton, NJ, 1984), p. 81.

98 Friedman, 'Architect's Compass in Creation Miniatures', p. 426. Recently, Humpert and Schenk, *Entdeckung der mittelalterlichen Stadtplanung*, pp. 319–23, have revealed the geometry embedded in the Vienna National Library Ms. 2554 drawing of God the artificer, demonstrating that it is itself produced by using a compass – geometry creating geometry.

99 For the architect's use of the compass/dividers see Shelby, 'Medieval Masons' Tools', though he has nothing to say on the cosmic shape inscribed by the instruments. On the Christian symbolism of the compass, and its use as a geometrical measuring instrument, see Zaitsev, 'Meaning of Early Medieval Geometry', p. 539.

100 Ezekiel 40–48, especially 40:3, 42:20.

101 Ezekiel 43:15–19; 48:15.

102 Ezekiel 48:16–17.

103 Victor, ed. and trans., *Practical Geometry in the High Middle Ages*, pp. 35, 37, citing Richard of St Victor, *In visionem Ezechielis*, and Roger Bacon, *Opus majus*.

104 See P. Grierson, *English Linear Measures an Essay in Origins* (Reading, 1972), pp. 20–24.

105 See above.

106 Revelation II.I; 21:15–16.

FIVE: MORAL TOPOGRAPHIES

1 *De Laude Cestrie* cited in D. M. Palliser, *Chester: Contemporary Descriptions by Residents and Visitors* (Chester, 1980), pp. 6–7.

2 *De Laude Cestrie* cited in ibid., pp. 6–7.

3 Augustine, *Enarrationes in Psalmum*, in M. S. Terry, trans., *The Sibylline Oracles, Translated from the Greek into English Blank Verse* (New York, 1899), p. 56, n. 30.

4 See V. F. Hopper, *Medieval Number Symbolism: Its Sources, Meaning, and Influence on Thought and Expression* (New York, 1969), pp. 31, 84. Also Terry, trans., *The Sibylline Oracles*, p. 56.

5 Romans 5:15; 1 Corinthians 15:21–22, 45.

6 See D. Woodward, 'Medieval *mappaemundi*', in *The History of Cartography*, vol. I, *Cartography in Prehistoric, Ancient and Medieval Europe and the Mediterranean*, ed. J. B. Harley and D. Woodward (Chicago, IL, 1987), pp. 334–42.

7 C. T. Davis, 'An Early Florentine Political Theorist: Fra Remigio de' Girolami', *Proceedings of the American Philosophical Society*, CIV (1960), pp. 662–76.

8 Cited in T. Dean, ed., *The Towns of Italy in the Later Middle Ages* (Manchester, 2000), pp. 156–7. Dean suggests this is Aristotelian in conception but it seems also to have a Calcidian resonance.

9 For a critical overview see S. Reynolds, 'Medieval Urban History and the History of Political Thought', *Urban History Yearbook* (1982), pp. 14–23.

10 L. T. Smith, ed., *The Maire of Bristowe is Kalendar*, Camden Society, new series, V (1872), p. 70.

11 Ibid., pp. 69, 70.

12 Ibid., p. 70.

13 Ibid., *Maire of Bristowe is Kalendar*, p. 74.

14 A. F. Leach, ed., *Beverley Town Documents*, Selden Society, XIV (London, 1900), p. 34.
15 M. James, 'Ritual, Drama and Social Body in the Late Medieval English Town', *Past and Present*, XCVIII (1983), pp. 3–29, at p. 10.
16 Leach, ed., *Beverley Town Documents*, p. 61.
17 K. Forhan, trans., 'Brunetto Latini, *The Book of Treasure*', in *Medieval Political Theory – A Reader: The Quest for the Body Politic, 1100–1400*, ed. C. J. Nederman and K. Langdon Forhan (London, 1993), pp. 71–96, at p. 71. See F. J. Carmody, ed., *Li Livres dou Tresor* (Berkeley, CA, 1948), pp. xxix, xxxi–xxxii. It has been said that 'although Latini also used Aristotle in his political writings, his ideas reflect more an influence of Roman authorities than Aristotle', Hamilton Bleakley, 'The Art of Ruling in Aquinas' *De Regimine Principum*', *History of Political Thought*, XX (1999), pp. 575–602, at p. 578, n. 23.
18 Forhan, trans., 'Brunetto Latini, *The Book of Treasure*', pp. 72–3.
19 H. T. Riley, ed. and trans., *Munimenta Gildhallæ Londoniensis; Liber Albus, Liber Custumarum, et Liber Horn*, Rolls Series, II, pt. 2 (1860), p. 517.
20 Reynolds, 'Medieval Urban History', p. 22.
21 J. Catto, 'Andrew Horn: Law and History in Fourteenth-Century England', in *The Writing of History in the Middle Ages: Essays presented to Richard William Southern*, ed. R.H.C. Davis and J. M. Wallace-Hadrill (Oxford, 1981), pp. 367–91, at p. 388.
22 Forhan, trans., 'Brunetto Latini, *The Book of Treasure*', p. 73.
23 Ibid., p.73.
24 Riley, ed. and trans., *Liber Custumarum*, p. 518.
25 Catto, 'Law and History in Fourteenth-Century England', p. 387. See also G. A. Williams, *Medieval London from Commune to Capital* (London, 1963), pp. 312–13; C. M. Barron, *London in the Later Middle Ages: Government and People 1200–1500* (Cambridge, 2004), p. 181.
26 Catto, 'Law and History in Fourteenth-Century England', p. 390.
27 Smith, ed., *Maire of Bristowe is Kalendar*, pp. xx–xxi, pp. 92–113, written in Old French 'for the most part in Ricart's own hand' (p. 92). Ricart says these came from when Henry Darcy was 'recorder of London', though Toulmin notes that Darcy was the city's mayor and had held office in 1337 and 1338 (p. xx).
28 Especially on 'the form in which the new mayor ought to be chosen' and how he should 'conduct and comport himself towards those whom he has in subjection under him, upon the last day of his holding office, when he has to take his leave of them', Riley, ed. and trans. *Liber Custumarum*, pp. 521–22. Compare with Smith, ed., *Maire of Bristowe is Kalendar*, pp. 71, 72: some of Ricart's other material in the *Kalendar* was taken from the *Liber Custumarum*, as well as London's *Liber Albus* (p. 93).
29 See, for example, Nederman and Langdon Forhan, ed., *Medieval Political Theory*, pp. 1–17.
30 Catto, 'Law and History in Fourteenth-Century England', p. 388. See J. H. Mundy, 'In Praise of Italy: The Italian Republics', *Speculum*, LXIV (1989), pp. 815–34, especially p. 816.
31 Hamilton Bleakley, 'Art of Ruling in Aquinas', pp. 575–602.
32 J. Catto, 'Ideas and Experience in the Political Thought of Aquinas', *Past and Present*, LXXI (1976), pp. 3–21, at pp. 13, 14.
33 Ibid., p. 19.
34 Hamilton Bleakley, 'Art of Ruling in Aquinas', p. 592, citing Aquinas, *De Regimine Principum*. Cf. J. M. Blythe, ed. and trans., *On the Government*

of Rulers, 'De Regimine Principum', Ptolemy of Lucca with Portions Attributed to Thomas Aquinas (Philadelphia, PA, 1997), pp. 96–7.

35 Hamilton Bleakley, 'Art of Ruling in Aquinas', p. 576, citing Aquinas, *In Octo Libros Politicorum Aristotelis Expositio*. Cf. S. Ziller, C. Nederman and K. Forhan, ed. and trans., 'Thomas of Aquinas, *On Kingship*, *Summary of Theology* and *Commentary on Aristotle's Politics*', in *Medieval Political Theory* ed. Nederman and Langdon Forhan (London, 1993), p. 137.

36 Blythe, ed. and trans., *On the Government of Rulers*, pp. 94, 97.

37 Hamilton Bleakley, 'Art of Ruling in Aquinas', pp. 593–4, citing Aquinas, *De Regimine Principum*. Cf. Blythe, ed. and trans., *On the Government of Rulers*, pp. 101–103.

38 Hamilton Bleakley, 'Art of Ruling in Aquinas', p. 576.

39 Catto, 'Ideas and Experience', p. 20; J. Dunbabin, 'Guido Vernani of Rimini's Commentary on Aristotle's *Politics*', *Traditio*, XLIV (1988); Ziller et al., ed. and trans., 'Thomas of Aquinas, *On Kingship*', p. 98.

40 C. J. Nederman, trans., 'Marsiglio of Padua, *The Defender of the Peace*', in *Medieval Political Theory*, ed. Nederman and Langdon Forhan, p. 173.

41 Ibid., p. 179; J. Canning, 'The Role of Power in the Political Thought of Marsilius of Padua', *History of Political Thought*, XX (1999), pp. 21–34, at p. 32, citing R. Scholz, ed., *Marsilius of Padua, Defensor pacis*, no. 7 in *MGH, Fontes iuris Germanici antique in usum scholarum* (Hanover, 1932), pp. 11–12.

42 Nederman, trans., 'Marsiglio of Padua, *The Defender of the Peace*', pp. 185, 186.

43 Ibid., p. 188.

44 Canning, 'Political Thought of Marsilius of Padua', p. 29.

45 C. J. Nederman, 'Nature, Sin and the Origins of Society: The Ciceronian Tradition in Medieval Political Thought', *Journal of the History of Ideas*, xl (1988), pp. 3–26.

46 Canning, 'Political Thought of Marsilius of Padua', p. 32.

47 Nederman, 'Nature, Sin and the Origins of Society', pp. 6–7.

48 Ibid., p. 23, see also p. 19.

49 Hamilton Bleakley, 'Art of Ruling in Aquinas', pp. 578–9, citing W. Miller, ed. *Cicero, De Officiis* (London, 1913), p. 87.

50 Blythe, ed. and trans., *Government of Rulers*, pp. 60–67, 271–4 (Book I, chapters 1–3; Book IV, chapter 23).

51 Hamilton Bleakley, 'Art of Ruling in Aquinas', p. 579, citing C. Salvemini, ed., *Giovanni da Viterbo, Liber de Regimine Civitatum*, in *Biblioteca juridical medii aevi*, III (Bologna 1901), pp. 220–21.

52 See Catto, 'Ideas and Experience'; Dunbabin, 'Guido Vernani of Rimini'; Canning, 'Political Thought of Marsilius of Padua'.

53 See K. Forhan, trans., 'Christine de Pizan, *The Book of the Body Politic*', in *Medieval Political Thought*, ed. Nederman and Langdon Forhan, pp. 230–47, eg. pp. 231–2.

54 C. J. Nederman, trans., 'John of Salisbury, *Metalogicon* and *Policraticus*', in *Medieval Political Thought*, ed. Nederman and Langdon Forhan, pp. 26–60, at p. 27.

55 S. H. Beer, 'The Rule of the Wise and the Holy: Hierarchy in the Thomistic System', *Political Theory*, XIV (1986), pp. 391–422, at p. 394.

56 Barron, *London in the Later Middle Ages*, pp. 147, 151–2; for a discussion of the procession see also L. J. Morrissey, 'English Pageant-wagons', *Eighteenth-Century Studies*, IX (1976), pp. 353–74.

57 For example, Winchester's 'usages' and Lincoln's 'provisions', see

H. Rothwell, ed., *English Historical Documents 1189–1327* (London, 1975), pp. 870–81.
58 For example, see Beer, 'Hierarchy in the Thomistic System', p. 399.
59 Reynolds, 'Medieval Urban History'.
60 See S. Reynolds, *Kingdoms and Communities in Western Europe, 900–1300* (Oxford, 1984).
61 A. J. Gurevich, *Categories of Medieval Culture*, trans. G. L. Campbell (London, 1985), p. 172.
62 See Introduction.
63 K. D. Lilley, '*Non urbe, non vico, non castris*: Territorial Control and the Colonization and Urbanization of Medieval Wales and Ireland under Anglo-Norman Lordship', *Journal of Historical Geography*, XXVI (2000), pp. 517–31.
64 R. Bartlett, *The Making of Europe: Conquest, Colonization and Cultural Change 950–1350* (London, 1993), pp. 172–82.
65 J. Campbell, 'Power and Authority, 600–1300', in *The Cambridge Urban History of Britain 600–1540*, ed. D. M. Palliser (Cambridge, 2000), pp. 51–69.
66 A. Williams and G. H. Martin, ed. and trans., *Domesday Book, a Complete Translation* (London, 2002), p. 688 (f. 252).
67 M. Chibnall, *Anglo-Norman England 1066–1166* (Oxford, 1986), pp. 13–19, 36.
68 M. Bateson, 'The Laws of Breteuil', *English Historical Review*, XV (1900), pp. 73–8, 302–18, 496–523, 754–7; *English Historical Review*, XVI (1901), pp. 92–110, 332–45.
69 J. Hillaby, 'The Norman New Town of Hereford: Its Street Pattern and European Context', *Transactions of the Woolhope Naturalists' Field Club*, XLIV (1982), pp. 181–95.
70 Williams and Martin, ed. and trans., *Domesday Book*, p. 493 (f. 179).
71 Bateson, 'Laws of Breteuil', pp. 311–14.
72 K. D. Lilley, *Urban Life in the Middle Ages, 1000–1450* (London, 2002), pp. 95–8.
73 K. D. Lilley, 'Mapping Cosmopolis: Moral Topographies of the Medieval City', *Environment and Planning d: Society and Space*, XXII (2004), pp. 681–98.
74 See Lilley, *Urban Life in the Middle Ages*, pp. 83–104.
75 D. H. Owen, 'Denbigh', in *Boroughs of Mediaeval Wales*, ed. R. A. Griffiths (Cardiff, 1978), pp. 164–87; R. I. Jack, 'Ruthin', in *Boroughs of Mediaeval Wales*, ed. Griffiths, pp. 244–61.
76 See Lilley, "*Non urbe, non vico, non castris*'.
77 Lilley, 'Mapping Cosmopolis'.
78 Ibid.
79 Ibid.
80 See above.
81 See Gurevich, *Categories of Medieval Culture*, pp. 61, 70–71.
82 K. R. Potter, ed. and trans., *Gesta Stephani* (Oxford, 1976), p. 57. R.H.C. Davis suggests the 'deeds of Stephen' was written possibly by Robert of Lewes, bishop of Bath, in the middle years of the twelfth century, Potter, ed. and trans., *Gesta Stephani*, pp. xx–xxi, xxxviii.
83 H. A. Cronne, ed., *Bristol Charters*, Bristol Records Society Publications I (Bristol, 1930), pp. 81–3.
84 Ibid, pp. 16, 25, 27.
85 Ibid, pp. 76–7. In 1344, 48 burgesses 'were elected to assist the mayor

in ensuring the better government of the town', p. 11.
86 Ibid, pp. 10–11; F. B. Bickley, ed., *Little Red Book of Bristol* (Bristol, 1900).
87 Ibid., II, p. 224.
88 Ibid., II, pp. 228–9. It is of note that these two proclamations (§§25 and 30) appear alongside others that ruled against rubbish in the streets (§§23 and 24) and cleaning 'the ways' in front of houses (§29). It is as if lepers and prostitutes were being viewed as part of the same detritus to be cleared from the town's streets.
89 Ibid., I, p. 33.
90 Leviticus 13:46.
91 Leviticus 14:45.
92 Ibid., *Little Red Book of Bristol*, I, pp. 33–34.
93 A. H. Smith, *Place-Names of Gloucestershire* (Cambridge, 1964), pp. 88, 92; F. B. Bickley, ed., *A Calendar of Deeds (Chiefly Relating to Bristol)* (Edinburgh, 1899), document 112, cf. 135, 143.
94 S. Seyer, ed., *Charters and Letters Patent Granted by the Kings and Queens of England to the Town and City of Bristol* (Bristol, 1812), p. 75.
95 E. O. Blake, ed., *The Cartulary of the Priory of St Denys near Southampton*, Southampton Records Series, XXIV (1981), p. lxvi; K. I. Sandred and B. Lingström, *The Place-Names of Norfolk, part I* (Irthingborough, 1989), p. 122.
96 B. Geremek, *The Margins of Society in Late Medieval Paris*, trans. J. Birrell (Cambridge, 1987).
97 C. Rawcliffe, 'The Earthly and Spiritual Topography of Suburban Hospitals', in *Town and Country in the Middle Ages*, ed. Giles and Dyer, pp. 251–74.

SIX: PERFORMING BODIES

1 M. James, 'Ritual, Drama and Social Body in the Late Medieval English Town', *Past and Present*, XCVIII (1983), pp. 3–29.
2 M. Eliade, *The Sacred and the Profane, the Nature of Religion*, trans. W. R. Trask (London, 1959), p. 105.
3 See E. Bishop, *Liturgica Historica: Papers on the Liturgy and Religious Life of the Western Church* (Oxford, 1918), especially pp. 276–300.
4 Eliade, *Sacred and Profane*, pp. 106–7.
5 J. F. Baldovin, *The Urban Character of Christian Worship in Jerusalem, Rome, and Constantinople from the Fourth to the Tenth Centuries: The Origins, Development, and Meaning of Stational Liturgy* (New Haven, CT, 1982), pp. 36–7, 234–8: stational liturgy 'was mobile: it did not always take place at the same church but was celebrated in different sanctuaries and shrines', the choice of which 'depended on the feast, fast, or commemoration being celebrated'.
6 Ibid., p. 251.
7 Ibid., p. 238. This classification deserves closer scrutiny. In processions of the later Middle Ages, such as those on the feast of Corpus Christi, both roles were surely being fulfilled. See M. Rubin, *Corpus Christi: The Eucharist in Late Medieval Culture* (Cambridge, 1991), pp. 347–61.
8 Baldovin, *Urban Character of Christian Worship*, p. 268.
9 James, 'Ritual, Drama and Social Body', pp. 3–29.
10 Ibid., p. 4.
11 G. Wickham, *Medieval Theatre* (Cambridge, 1987), p. 63. Rubin, *Corpus*

Christi, pp. 172–81: she notes (p. 179) 'the diffusion of the feast down to 1300 was stimulated by initiatives of particular bishops, like [William] Durandus, in their areas of influence, but there was clearly no driving force from Rome'.

12 James, 'Ritual, Drama and Social Body', p. 4.

13 Ibid., p. 6. See also G. Nijsten, 'Feasts and Public Spectacle: Late Medieval Drama and Performance in the Low Countries', in A. E. Knight, ed., *The Stage as Mirror: Civic Theatre in Late Medieval Europe* (Cambridge, 1997), pp. 107–44, especially pp. 126–8.

14 See Introduction.

15 James, 'Ritual, Drama and Social Body', p. 9.

16 Matthew 26:26; cf. Mark 14:22, Luke 22:19. On the institution of the feast day see V. A. Kolve, *The Play Called Corpus Christi* (Stanford, CA, 1966), pp. 43–46; Rubin, *Corpus Christi*, pp. 178–9.

17 James, 'Ritual, Drama and Social Body', p. 9.

18 For example, in Beverley and London: see chapter Five. Also James, 'Ritual, Drama and Social Body', p. 11.

19 For example, with the arrival of Richard III in 1483 in York 'displays were erected at the gate by which Richard entered the city and on major streets as he made his way on horseback through York', the 'royal procession trod the same path as the city's annual Corpus Christi play cycle', and 'although he had missed their performance three months earlier, Richard had timed his arrival to a date particularly hallowed in the city . . . 29 August, the feast of the Decollation of John the Baptist, a symbol the citizens understood to be analogous to the Body of Christ worshipped so visibly by their leading guild and the ceremonies it sponsored', L. Atreed, 'The Politics of Welcome: Ceremonies and Constitutional Development of Later Medieval English Towns', in *City and Spectacle in Medieval Europe*, ed. B. Hanawalt and K. Reyerson (Minneapolis, MN, 1994), pp. 208–31, at p. 217.

20 The following discussion is based upon this vast reservoir of literature on Corpus Christi processions and pageants, but for a sense of the two bodies of literature see especially (for histories of drama): M. Spencer, *Corpus Christi Pageants in England* (New York, 1911); Kolve, *Play Called Corpus Christi*; A. H. Nelson, *The Medieval English Stage: Corpus Christi Pageants and Plays* (Chicago, IL, 1974); L. R. Muir, *The Biblical Drama of Medieval Europe* (Cambridge, 1995); a particularly clear historiographical overview is D. Staines, 'The English Mystery Cycles', in *The Theatre of Medieval Europe: New Research in Early Drama*, ed. E. Simon (Cambridge, 1991), pp. 80–96; (and for cultural/social histories) C. Phythian Adams, 'Ceremony and the Citizen: The Communal Year at Coventry 1450–1550', in *Crisis and Order in English Towns, 1500–1700*, ed. P. Clark and P. Slack (London, 1972), pp. 57–85; James, 'Ritual, Drama and Social Body'; P. W. Travis, 'The Social Body of the Dramatic Christ in Medieval England', *Early English Drama: Acta*, XIII, pp. 17–36; Rubin, *Corpus Christi*; S. Beckwith, *Signifying God: Social Relation and Symbolic Act in the York Corpus Christi Plays* (Chicago, IL, 2001).

21 James, 'Ritual, Drama and Social Body', pp. 11, 12. Not all seem to agree with this view, however. Those who have drawn upon James's thesis have begun to doubt that the Corpus Christi procession held the city together by 'imparting to the wholeness of the social body an ordered structure which it would otherwise lack.' A developing critique of James's thesis has taken issue with him in two main respects: first is

his assumption that the body is a unifying metaphor. However, the body can mean different things to different people, and thus is not always perceived as a unifying whole but rather a site of conflict and differentiation. 'Bodies', Rubin (*Corpus Christi*, p. 270) thus explains, 'cannot be taken as possessing an essential meaning; like all meaningful signs they are culturally constructed' and 'in a political discourse, the invention of their naturalness [through divine archetype] in particular contexts may or may not be observed by those who observe it.' Similarly, Beckwith (*Signifying God*, p. 30) points out 'that Christ's body alludes to numerous oppositions: inner and outer, transcendent and imminent, spirit and flesh, male and female, left and right, up and down, noisy and silent, just and unjust, passive and active, noumenal and phenomenal, public and private, hierarchical and collective, unified and multiplicitous, and so on', all of which makes any one (civic/political) interpretation of *corpus Christi* begin to appear questionable. It is from this questioning of the meaning 'body' had that Rubin sees the body as contested ground – something fought over – and hence potentially a problematic model for medieval civic bodies to use. However, questioning James's thesis in terms of his views of medieval urban society and ritual does not diminish or undermine the view that performing *corpus Christi* embodied the social fabric of the city itself, and while not necessarily harmonious and ordered, nevertheless projected an idealized image of the city within and beyond the urban area, hence *symbolically* unifying the whole of the sum parts to those who were on the 'inside' and 'outside' of the urban body.

22 Nijsten, 'Feasts and Public Spectacle', p. 127. As a metaphor for the urban body, the Body of Christ was not without conflicting and contradictory connotations, and the civic ritual performance of processions celebrating *corpus Christi* likewise not without their socially destabilizing and opposing elements. In elaborating on the 'body' in these ways, Rubin and others have argued that social cohesion and 'wholeness' were by no means certain outcomes of Corpus Christi celebrations and civic display, and advocate instead a more nuanced approach to interpreting the body metaphor in the context of ritual performance. Thus for Rubin the Corpus Christi procession, 'even seen as a metaphoric body, will contain members in disagreement as to their appropriate hierarchical position as to who was the stomach, who the head, for how long, and why', and so (*contra* Phythian Adams), 'was not a picture of the community' but could lay 'hierarchy bare' and 'incite the conflict of difference ever more powerfully sensed in a concentrated symbolic moment'. Rubin, *Corpus Christi*, pp. 270–71, 266, cf. C. Phythian Adams, *Desolation of a City: Coventry and the Urban Crisis of the Late Middle Ages* (Cambridge, 1979), p. 178. See also S. Lindenbaum, 'Ceremony and Oligarchy: The London Midsummer Watch', in *City and Spectacle in Medieval Europe*, ed. Hanawalt and Reyerson, pp. 171–88; and B. R. McRee, 'Unity or Division? The Social Meaning of Guild Ceremony in Urban Communities', in *City and Spectacle in Medieval Europe*, ed. Hanawalt and Reyerson, pp. 189–207.

23 This is Nijsten's view, see 'Feasts and Public Spectacle'. See also R. B. Dobson, 'Craft Guilds and City: The Historical Origins of the York Mystery Plays Reassessed', in *Stage as Mirror*, ed. Knight, pp. 91–105. How far processions' participants believed in this civic rhetoric is obviously difficult to judge. Of course these collective, civic expressions of

piety were reflecting and reinforcing social difference and urban order, and yes, as Rubin (*Corpus Christi*, p. 266) notes, 'a procession which excluded most working people, women, children, visitors, and servants, was not a picture of the community', but this should not necessarily be taken to mean that such events were seen as exclusive and exclusionary.

24 See W. Tydeman, *The Theatre in the Middle Ages* (Cambridge, 1978), pp. 184–221.

25 Beckwith, *Signifying God*, p. 28.

26 James, 'Ritual, Drama and Social Body', pp. 6–7: 'natural body and social body indeed reacted on each other with a closeness which comes to near-identity'.

27 Rubin, *Corpus Christi*, p. 270: 'to the natural symbol of the well functioning and harmonious body, they added the symbol of Christ's body, which is both identiical with and separate from the social body, and much celebrated for its power'.

28 Nijsten, 'Feasts and Public Spectacle', pp. 127–8; T. A. Boogaart II, 'Our Saviour's Blood: Procession and Community in Late Medieval Bruges', in *Moving Subjects: Processional Performance in the Middle Ages and the Renaissance*, ed. K. Ashley and W. Hüskin (Amsterdam, 2001), pp. 69–116.

29 For a useful summary see Rubin, *Corpus Christi*, pp. 267–9.

30 For example, see papers in Knight, ed., *Stage as Mirror*; Simon, ed., *Theatre of Medieval Europe*.

31 A point made but not developed by B. Hanawalt and K. Reyerson, 'Introduction', in *City and Spectacle in Medieval Europe*, ed. Hanawalt and Reyerson, p. xviii.

32 K. Ashley, 'Introduction: The Moving Subjects of Processional Performance', in *Moving Subjects*, ed. Ashley and Hüskin, p. 17.

33 Boogaart, 'Procession and Community in Late Medieval Bruges', p. 94.

34 Ibid., p. 70. For the development of Bruges and its place in medieval Flanders, see M. Ryckaert, *Historische Stedenatlas van België. Brugge* (Brussels, 1991), pp. 66–107; also, A. Brown, 'Civic Ritual: Bruges and the Counts of Flanders in the Later Middle Ages', *English Historical Review*, XXII (1997), pp. 277–99.

35 Boogaart, 'Procession and Community in Late Medieval Bruges', pp. 74, 75.

36 Ibid., p. 87.

37 Ibid., p. 88.

38 Ibid., p. 90.

39 Ibid., pp. 88–91.

40 Ibid., p. 89 and see n. 145. The procession also 'involved the ritual usage of processional crosses at cardinal points' (p. 100, n. 25).

41 Ibid., p. 89.

42 See Ashley, 'Moving Subjects of Processional Performance', pp. 17–18.

43 C. Zika, 'Hosts, Processions and Pilgrimages: Controlling the Sacred in Fifteenth-Century Germany', *Past and Present*, XVIII (1988), pp. 25–64, at pp. 38–9.

44 Ibid., p. 39.

45 Ibid., p. 39. See Boogaart, 'Procession and Community in Late Medieval Bruges', p. 75 and n. 40.

46 See A. Haverkamp, 'Heilige Städte im hohen Mittelater', in *Mentalitäten im Mittelalter: Methodische und inhaltliche Probleme*, ed. F. Grau (Sigmaringen, 1987), pp. 119–56, cited in Boogaart, 'Procession and Community', note 150.

47 K. J. Allison, ed., *A History of the County of York East Riding*, vol. XI: *The Borough and Liberties of Beverley* (London, 1989), pp. 42–9; L. Blair, 'A Note on the Relation of the Corpus Christi Procession to the Corpus Christi Play in England', *Modern Language Notes*, LV (1940), pp. 83–95; Nelson, *Corpus Christi Pageants and Plays*, pp. 38–207.
48 A. F. Leach, ed., *Beverley Town Documents*, Selden Society, XIV (London, 1900), p. 34.
49 James, 'Ritual, Drama and Social Body', p. 10. See also chapter Five.
50 Nelson, *Corpus Christi Pageants and Plays*, pp. 90–97.
51 A. F. Leach, ed., *Report on the Manuscripts of the Corporation of Beverley* (London, 1900), p. 133; Allison, ed., *Borough and Liberties of Beverley*, p. 47.
52 Nelson, *Corpus Christi Pageants and Plays*, p. 97.
53 For example, see Wickham, *Medieval Theatre*; Kolve, *Play Called Corpus Christi*; Nelson, *Corpus Christi Pageants and Plays.*
54 L. Powlick, 'The Staging of the Chester Cycle: An Alternate Theory', in *The Chester Mystery Cycle: A Casebook*, ed. K. J. Harty (London, 1993), pp. 199–229, especially pp. 208–19. Other routes are also recorded, see L. M. Clopper, ed., *Chester, Records of Early English Drama* (London, 1979), pp. liii–lviii; L. M. Clopper, 'The History and Development of the Chester Cycle', *Modern Philology*, LXXV (1978), pp. 219–46.
55 There is some uncertainty over the exact dramaturgical and temporal relationship between the Corpus Christi plays at Chester and the city's 'Whitsun Plays', see Clopper, ed., *Chester*, pp. liii–liv; Clopper, 'History and Development', p. 221; Powlick, 'Staging of the Chester Cycle', p. 208.
56 See Nelson, *Corpus Christi Pageants and Plays*, pp. 38–45; A. F. Johnston, 'The Guild of Corpus Christi and the Procession of Corpus Christi in York', *Mediaeval Studies*, XXXVIII (1976), pp. 372–84.
57 Nelson, *Corpus Christi Pageants and Plays*, pp. 39, 45.
58 J. F. Hoy, 'On the Relationship of the Corpus Christi Plays to the Corpus Christi Procession at York', *Modern Philology*, LXXI (1973), p. 166, citing M. Sellers, ed., *York Memorandum, Book 2* (Durham, 1912–15), f.19b. See A. Higgins, 'Streets and Markets', in *A New History of Early English Drama*, ed. J. D. Cox and D. S. Kastan (New York, 1997), pp. 77–92.
59 Hoy, 'Relationship of the Corpus Christi Plays', p. 167, citing Sellers, ed., *York Memorandum Book 2*, f.278.
60 Compare Nelson, *Corpus Christi Pageants and Plays*, pp. 15–37; S. J. Karhl, 'The Staging of Medieval English Plays', in *Theatre of Medieval Europe*, ed. Simon, pp. 130–48, at pp. 138–9.
61 E. White, 'Places for Hearing the Corpus Christi Play in York', *English Medieval Theatre*, IX (1987), pp. 23–63.
62 A. J. Mill, 'The Stations of the York Corpus Christi Play', *Yorkshire Archaeological Journal*, XXXVII (1951), pp. 492–502.
63 See P. Ottaway, *Archaeology in British Towns: From the Emperor Claudius to the Black Death* (London, 1992), pp. 82–95, 146–9, fig. 5.9 on p. 147.
64 See also Higgins, 'Streets and Markets', pp. 83–6.
65 H. Craig, ed., *Two Coventry Corpus Christi Plays: 1. The Shearmen and Taylors' Pageant, and 2. The Weavers' Pageant* (London, 1957), second edn, pp. xiii–xiv; W. B. Stephens, ed., *A History of the County of Warwick*, vol. VIII: *The City of Coventry and Borough of Warwick* (London, 1969), p. 214.
66 N. Coulet, 'Processions, espace urbain, communauté civique', *Cahiers*

de Fanjeaux, XVII (1982), pp. 381–97.
67 See Rubin, *Corpus Christi*, p. 265; Boogaart, 'Procession and Community in Late Medieval Bruges', p. 94.
68 Boogaart, 'Procession and Community in Late Medieval Bruges', p. 90.
69 Rubin, *Corpus Christi*, p. 271.
70 Muir, *Biblical Drama of Medieval Europe*, p. 22; Tydeman, *Theatre in the Middle Ages*, pp. 97–120.
71 See Karhl, 'Staging of Medieval English Plays', pp. 130–48.
72 Muir, *Biblical Drama of Medieval Europe*, pp. 24–5.
73 Staines, 'English Mystery Cycles', p. 80: in 'English dramatic criticism' the term 'cycle' was coined by Lucy Toulmin Smith in 1884 (p. 83).
74 Translated and cited in Rubin, *Corpus Christi*, p. 277.
75 Nijsten, 'Feasts and Public Spectacle', pp. 126, 127.
76 S. J. Karhl, *Traditions of Medieval English Drama* (London, 1974), p. 124.
77 Kolve, *Play Called Corpus Christi*, p. 42; Staines, 'English Mystery Cycles', p. 80.
78 L. T. Smith, ed. *York Plays: The Plays Performed by the Crafts or Mysteries of York on the Day of Corpus Christi* (Oxford, 1885), pp. lxii–lxiii.
79 Smith, ed. *York Plays*, pp. lxii–lxiii.
80 Mill, 'Stations of the York Corpus Christi Play', appendix 3.
81 Ibid., appendix 3.
82 Smith, ed., *York Plays*, p. xxxii.
83 See Nelson, *Corpus Christi Pageants and Plays*, pp. 15–37.
84 Tydeman, *Theatre in the Middle Ages*, p. 117, also pp. 112–17; also Nelson, *Corpus Christi Pageants and Plays*, pp. 15–37; Karhl, 'Staging of Medieval English Plays', pp. 138–9.
85 Nelson, *Corpus Christi Pageants and Plays*, p. 33.
86 H. Craig, *English Religious Drama of the Middle Ages* (Oxford, 1955), p. 133.
87 J. Taylor, 'The Dramatic Structure of the Middle English Corpus Christi, or Cycle, Plays', in *Medieval English Drama: Essays Critical and Contextual*, ed. J. Taylor and A. H. Nelson (London, 1972), pp. 148–56, at p. 149.
88 Nijsten, 'Feasts and Public Spectacle', p. 128.
89 Karhl, *Traditions of Medieval English Drama*, p. 125.
90 Higgins, 'Streets and Markets', p. 91.
91 Rubin, *Corpus Christi*, p. 278.
92 See Nijsten, 'Feasts and Public Spectacle'; Muir, *Biblical Drama of Medieval Europe*; papers on France, Italy, Spain and the Low Countries in Simon, ed., *Theatre of Medieval Europe*, pp. 151–252, with extensive bibliography (pp. 266–90).
93 Boogaart, 'Procession and Community in Late Medieval Bruges', p. 93.
94 Ibid., pp. 76, 90.
95 Rubin, *Corpus Christi*, pp. 243–71.
96 Boogaart, 'Procession and Community in Late Medieval Bruges', p. 94.
97 B. Scribner, 'Cosmic Order and Daily Life: Sacred and Secular in Pre-industrial German Society', in *Religion and Society in Early Modern Europe 1500–1800*, ed. K. von Greyerz (London, 1984), pp. 17–32, at p. 18.
98 Boogaart, 'Procession and Community in Late Medieval Bruges', pp. 74–5, see also pp. 83, 90; see also Brown, 'Bruges and the Counts of Flanders', p. 281.
99 Powlick, 'Staging of the Chester Cycle', pp. 218–19.
100 Clopper, 'History and Development', pp. 243–4.
101 Powlick, 'Staging of the Chester Cycle', pp. 221–7.

102 K. M. Ashley, 'Divine Power in Chester Cycle and Late Medieval Thought', in *Chester Mystery Cycle*, ed. Harty, pp. 48–68, at pp. 57, 61.
103 A. M. Nagler, *The Medieval Religious Stage: Shapes and Phantoms* (London, 1976), p. 16.
104 By Julius Petersen, see Nagler, *Medieval Religious Stage*, pp. 16–17.
105 Nagler, *Medieval Religious Stage*, pp. 17–18.
106 P. Meredith and J. E. Tailby, ed., *The Staging of Religious Drama in Europe in the Later Middle Ages: Texts and Documents in Translation* (Kalamazoo, MI, 1983), p. 83; also Nagler, *Medieval Religious Stage*, pp. 29–32.
107 Higgins, 'Streets and Markets', p. 91.
108 Smith, ed., *York Plays*, p. xxxiii; Mill, 'Stations of the York Corpus Christi Play', p. 493.
109 Cited in M. Twycross, 'Some Approaches to Dramatic Festivity, Especially Processions', in *Festive Drama*, ed. M. Twycross (Cambridge, 1996), pp. 1–33, at p. 4.
110 Nijsten, 'Feasts and Public Spectacle', p. 126.

EPILOGUE: CITIES OF GOD?

1 K. D. Lilley, 'Cities of God? Medieval Urban Forms and their Christian Symbolism', *Transactions of the Institute of British Geographers*, new series, XXIX (2004), pp. 296–313.
2 See G. P. Klein, 'The Topography of Symbol: Between Late Antique and Modern Jewish Understanding of Cities', *Zeitschrift für Religions- und Geistesgeschichte*, LVIII (2006), pp. 16–28.
3 T. J. Gorringe, *A Theology of the Built Environment: Justice, Empowerment, Redemption* (Cambridge, 2002).
4 P. V. Begde, *Ancient and Medieval Town Planning in India* (New Delhi, 1978), pp. 28–49.
5 See ibid., pp. 51–111; H.-J. Nitz, 'Planned Temple Towns and Brahmin Villages as Spatial Expressions of the Ritual Politics of Medieval Kingdoms in South India', in *Ideology and Landscape in Historical Perspective*, ed. A. Baker and G. Biger (Cambridge, 1992), pp. 107–24, at p. 117; P. Wheatley, *The Pivot of the Four Quarters: a Preliminary Enquiry into the Origins and Character of the Ancient Chinese City* (Edinburgh, 1971); N. Steinhardt, *Chinese Imperial City Planning* (Honolulu, 1992); A. F. Wright, 'The Cosmology of the Chinese City', in *The City in Late Imperial China*, ed. G. W. Skinner (Stanford, CA, 1977), pp. 33–73; A. Southall, 'The Circle and the Square: Symbolic Form and Process in the City', in *Urban Symbolism, Leiden Studies in Human Society*, ed. P.M.J. Nas (Leiden, 1993), pp. 378–93.
6 J. Rykwert, *The Idea of a Town: The Anthropology of Urban Form in Rome, Italy and the Ancient World* (London, 1988), pp. 163–93.
7 P. Wheatley, *City as Symbol* (London, 1969).
8 R. I. Levy, *Mesocosm: Hinduism and the Organisation of a Traditional Newar City in Nepal* (Berkeley, CA, 1990), p. 403; T. A. Boogaart II, 'Our Saviour's Blood: Procession and Community in Late Medieval Bruges', in *Moving Subjects: Processional Performance in the Middle Ages and the Renaissance*, ed. K. Ashley and W. Hüskin (Amsterdam, 2001), pp. 94–6.

SELECT BIBLIOGRAPHY

Alexander, J.J.G., '"Jerusalem the Golden": Image and Myth in the Middle Ages in Western Europe', in *The Real and Ideal Jerusalem in Jewish, Christian and Islamic Art*, ed. B. Kühnel (Jerusalem, 1998), pp. 254–64

Allers, R., 'Microcosmus: From Anaximandros to Paracelsus', *Traditio*, II (1944), pp. 319–407

Allison, K. J., ed., *A History of the County of York East Riding*, vol. XI: *The Borough and Liberties of Beverley* (London, 1989)

Ashley, K., 'Introduction: The Moving Subjects of Processional Performance', in *Moving Subjects: Processional Performance in the Middle Ages and the Renaissance*, ed. K. Ashley and W. Hüskin (Amsterdam, 2001), pp. 7–34.

—, and W. Hüskin, ed., *Moving Subjects: Processional Performance in the Middle Ages and the Renaissance* (Amsterdam, 2001)

Ashley, K. M., 'Divine Power in Chester Cycle and Late Medieval Thought', in *The Chester Mystery Cycle: A Casebook*, ed. K. J. Harty (London, 1993), pp. 48–68

Aston, M., and C. J. Bond, *The Landscape of Towns* (London, 1976)

Atreed, L., 'The Politics of Welcome: Ceremonies and Constitutional Development of Later Medieval English Towns', in *City and Spectacle in Medieval Europe*, ed. B. Hanawalt and K. Reyerson (Minneapolis, MN, 1994), pp. 208–31

Bachrach, B. S., 'Imperial Walled Cities in the West: An Examination of their Early Medieval Nachleben', in *City Walls: The Urban Enceinte in Global Perspective*, ed. J. D. Tracy (Cambridge, 2000), pp. 192–218

Baker, A., and G. Biger, ed., *Ideology and Landscape in Historical Perspective* (Cambridge, 1992)

Baldovin, J. F., *The Urban Character of Christian Worship in Jerusalem, Rome, and Constantinople from the Fourth to the Tenth Centuries: The Origins, Development, and Meaning of Stational Liturgy* (New Haven, CT, 1982)

Baron, R., 'Hughes de Saint-Victor, auteur d'une *Practica Geometriae*', *Mediaeval Studies*, XVII (1955), pp. 107–16

—, 'Sur l'introduction en Occident des termes "geometria theorica et practica"', *Revue d'histoire des sciences et de leurs applications*, VIII (1955), pp. 298–302

Barron, C. M., *London in the Later Middle Ages: Government and People 1200–1500* (Cambridge, 2004)

Barrow, R. H., *Introduction to St Augustine, 'The City of God'* (London, 1950)

Bartlett, R., *The Making of Europe: Conquest, Colonization and Cultural Change 950–1350* (London, 1993)

Bascapé, G. C., *Sigillografia. Il digillo nella diplomatica, nel diritto, nella storia, nell' arte* (Milan, 1969)

Bateson, M., 'The Laws of Breteuil', *English Historical Review*, XV (1900), pp. 73–8, 302–18, 496–523, 754–7; XVI (1901), pp. 92–110, 332–45

Baur, L., ed., *Dominicus Gundissalinus, 'De divisione philosophiae'* (Münster, 1903)

Beckwith, S., *Signifying God: Social Relation and Symbolic Act in the York Corpus Christi Plays* (Chicago, IL, 2001)
Beer, S. H., 'The Rule of the Wise and the Holy: Hierarchy in the Thomistic System', *Political Theory*, XIV (1986), pp. 391–422
Begde, P. V., *Ancient and Medieval Town Planning in India* (New Delhi, 1978)
Benevolo, L., *The European City*, trans. C. Ipsen (Oxford, 1993)
Beresford, M. W., *New Towns of the Middle Ages: Town Plantation in England, Wales and Gascony* (London, 1967)
Bettenson, H., ed. and trans., *Augustine, The City of God* (London, 1984)
Bickley, F. B., ed., *A Calendar of Deeds (Chiefly Relating to Bristol)* (Edinburgh, 1899)
—, ed., *Little Red Book of Bristol* (Bristol, 1900), 2 vols.
Biddle, M., and D. Hill, 'Late Saxon Planned Towns', *Antiquaries Journal*, LI(1971), pp. 70–85
Binding, G., *Medieval Building Techniques*, trans. A. Cameron (Stroud, 2004)
Bishop, E., *Liturgica Historica: Papers on the Liturgy and Religious Life of the Western Church* (Oxford, 1918)
Blacker, C., and M. Loewe, ed., *Ancient Cosmologies* (London, 1975)
Blackwell, R. J., R. J. Spath and W. E. Thirkel, trans., *Commentary on Aristotle's 'Physics': St Thomas Aquinas* (Notre Dame, IN, 1999)
Blair, L., 'A Note on the Relation of the Corpus Christi Procession to the Corpus Christi Play in England', *Modern Language Notes*, LV (1940) (1940), pp. 83–95
Blake, E. O., ed., *The Cartulary of the Priory of St Denys near Southampton*, Southampton Records Series, XXIV (1981)
Blythe, J. M., ed. and trans., *On the Government of Rulers, 'De Regimine Principum', Ptolemy of Lucca with portions attributed to Thomas Aquinas* (Philadelphia, PA, 1997)
Boerefijn, W., 'Designing the Medieval New Town', *Urban Morphology*, IV (2000), pp. 49–62
Boogaart II, T. A., 'Our Saviour's Blood: Procession and Community in Late Medieval Bruges', in *Moving Subjects: Processional Performance in the Middle Ages and the Renaissance*, ed. K. Ashley and W. Hüskin (Amsterdam, 2001), pp. 69–116
Bowsky, W. M., *A Medieval Italian Commune: Siena under the Nine, 1287–1355* (Berkeley, CA, 1981)
Branner, R., 'Villard de Honnecourt, Reims and the Origin of Gothic Architectural Drawing', *Gazette des Beaux-Arts*, LXI (1963), pp. 129–46
Brieger, P. H., *The Trinity College Apocalypse: An Introduction and Description* (London, 1967)
Brogiolo, G. P., and B. Ward-Perkins, ed., *The Idea and Ideal of the Town between Late Antiquity and the Early Middle Ages* (Leiden, 1999)
Brown, A., 'Civic Ritual: Bruges and the Counts of Flanders in the Later Middle Ages', *English Historical Review*, CXII (1997), pp. 277–99
Brown, R. A., H. M. Colvin and A. J. Taylor, *History of the King's Works*, vols I and II: *The Middle Ages* (London, 1963)
Bubnov, N., ed., *Gerberti Opera mathematica (972–1003)* (Berolini, 1899)
Bucher, F., *Architector: The Lodgebooks and Sketchbooks of Medieval Architects*, vol. I (New York, 1979)
—, 'Medieval Architectural Design Methods, 800–1560', *Gesta*, XI (1972), pp. 37–43
Bulmer Thomas, I., 'Euclid and Medieval Architecture', *Archaeological Journal*, CXXXVI (1979), pp. 136–50

Burnett, C., ed., *Adelard of Bath: An English Scientist and Arabist of the Early Twelfth Century* (London, 1987)
Busard, H.L.L., ed., *The First Translation of Euclid's 'Elements', Commonly Ascribed to Adelard of Bath* (Toronto, 1983)
—, ed., 'The *Practica Geometriae* of Dominicus de Clavasio', *Archive for the History of Exact Sciences*, II (1965), pp. 520–75
Butler, L., '"The Monastic City" in Wales: Myth or Reality?', *Bulletin of the Board of Celtic Studies*, XXVIII (1979), pp. 458–67
Callus, D. A., ed., *Robert Grosseteste: Scholar and Bishop* (Oxford, 1955)
—, 'Robert Grosseteste as Scholar', in *Robert Grosseteste: Scholar and Bishop*, ed. D. A. Callus (Oxford, 1955), pp. 1–69
Campbell, B., 'Shaping the Rural Environment: Surveyors in Ancient Rome', *Journal of Roman Studies*, LXXXVI (1996), pp. 74–99
—, ed. and trans., *The Writings of the Roman Land Surveyors: Introduction, Text, Translation and Commentary* (London, 2000)
Campbell, J., 'Power and Authority, 600–1300', in *The Cambridge Urban History of Britain*, vol. I: *600–1540*, ed. D. M. Palliser (Cambridge, 2000), pp. 51–69
Canning, J., 'The Role of Power in the Political Thought of Marsilius of Padua', *History of Political Thought*, XX (1999), pp. 21–34
Carder, J. N., *Art Historical Problems of a Roman Land Surveying Manuscript: The Codex Arcerianus A, Wolfenbüttel* (New York, 1978)
Carmody, F. J., ed., *Li Livres dou Tresor* (Berkeley, CA, 1948)
Catto, J., 'Andrew Horn: Law and History in Fourteenth-century England', in *The Writing of History in the Middle Ages: Essays Presented to Richard William Southern* (Oxford, 1981), ed. R.H.C. Davis and J. M. Wallace-Hadrill, pp. 367–91
—, 'Ideas and Experience in the Political Thought of Aquinas', *Past and Present*, LXXI (1976), pp. 3–21
Chenu, M. D., *Nature, Man, Society in the Twelfth Century: Essays on New Theological Perspectives in the Latin West*, ed. and trans. J. Taylor and L. K. Little (Chicago, IL, 1968)
Chibnall, M., *Anglo-Norman England 1066–1166* (Oxford, 1986)
Clark, J., 'King Alfred's London and London's King Alfred', *London Archaeologist*, IX (1999), pp. 35–8
Clark, P., and P. Slack, ed., *Crisis and Order in English Towns, 1500–1700* (London, 1972)
Clarke, C.A.M., *Literary Landscapes and the Idea of England, 700–1400* (Cambridge, 2006)
Clarke, H. B., and A. Simms, ed., *The Comparative History of Urban Origins in Non-Roman Europe* (Oxford, 1985)
—, 'Towards a Comparative History of Urban Origins', in *The Comparative History of Urban Origins in Non-Roman Europe* (Oxford, 1985), ed. H. B. Clarke and A. Simms, pp. 547–677
Clopper, L. M., ed., *Chester, Records of Early English Drama* (London, 1979)
—, 'The History and Development of the Chester Cycle', *Modern Philology*, LXXV (1978), pp. 219–46
Colson, F. H., and G. H. Whitaker, trans., *Philo* (Cambridge, MA, 1929)
Conger, G. P., *Theories of Macrocosms and Microcosms in the History of Philosophy* (New York, 1922)
Coulet, N., 'Processions, espace urbain, communauté civique', *Cahiers de Fanjeaux*, XVII (1982), pp. 381–97
Cox, J. D., and D. S. Kastan, ed., *A New History of Early English Drama* (New York, 1997)

Craig, H., *English Religious Drama of the Middle Ages* (Oxford, 1955)
—, ed., *Two Coventry Corpus Christi Plays: 1. The Shearmen and Taylors' Pageant, and 2. The Weavers' Pageant* (London, 1957), second edn
Cronne, H. A., ed., *Bristol Charters*, Bristol Records Society Publications I (Bristol, 1930)
Crouse, R. D., '*Hic sensilus mundus*: Calcidius and Eriugena in Honorius Augustudunensis', in *From Athens to Chartres: Neoplatonism and Medieval Thought: Studies in Honour of Edouard Jeauneau*, ed. H. J. Westra (Leiden, 1992), pp. 283–88
Crummy, P., 'The System of Measurement Used in Town Planning from the Ninth to the Thirteenth Centuries', in *Anglo-Saxon Studies in Archaeology and History*, vol. I, ed. S. C. Hawkes, D. Brown and J. Campbell (Oxford, 1979), pp. 149–64
Curie-Seimbres, A., *Essai sur les Villes Fondées dans le Sud-Ouest de la France aux XIIIe et XIVe Siècles sous le Nom Générique de Bastides* (Toulouse, 1880)
Davies, B., 'Thomas Aquinas', in *Routledge History of Philosophy*, vol. III: *Medieval Philosophy* (London, 1988), ed. J. Marenbon, pp. 241–68.
Davis, C. T., 'An Early Florentine Political Theorist: Fra Remigio de' Girolami', *Proceedings of the American Philosophical Society*, LIV (1960), pp. 662–76
Davis, R.H.C., 'An Oxford Charter of 1191 and the Beginnings of Municipal Freedom', *Oxoniensia*, XXXIII (1968), pp. 55–65
—, and J. M. Wallace-Hadrill, ed., *The Writing of History in the Middle Ages: Essays Presented to Richard William Southern* (Oxford, 1981)
Dean, T., ed., *The Towns of Italy in the Later Middle Ages* (Manchester, 2000)
Dilke, O.A.W., 'Illustrations from Roman Surveyors' Manuals', *Imago Mundi*, XXI (1967), pp. 9–29
Divorne, F., B. Gendre, B. Lavergne and P. Panerai, *Les Bastides d'Aquitaine, du Bas-Languedoc et du Béarn, Essai sur la Régularité* (Brussels, 1985)
Dobson, R. B., 'Craft Guilds and City: The Historical Origins of the York Mystery Plays Reassessed', in *The Stage as Mirror: Civic Theatre in Late Medieval Europe* (Cambridge, 1997), ed. A. E. Knight, pp. 91–105
Doherty, C., 'The Monastic Town in Early Medieval Ireland', in *The Comparative History of Urban Origins in Non-Roman Europe* (Oxford, 1985), ed. H. B. Clarke and A. Simms, pp. 45–75
Drew, A., 'The *De Eodem et Diverso*', in *Adelard of Bath: An English Scientist and Arabist of the Early Twelfth Century* (London, 1987), ed. C. Burnett, pp. 17–24
Dreyer, J.L.E., *History of the Planetary Systems from Thales to Kepler* (Cambridge, 1906)
Dronke, P., *Fabula: Explorations into the Uses of Myth in Medieval Platonism* (Leiden, 1974)
—, ed., *A History of Twelfth-Century Western Philosophy* (Cambridge, 1988)
—, 'Introduction', in *A History of Twelfth-Century Western Philosophy*, ed. P. Dronke (Cambridge, 1988), pp. 1–18
Drös, H., and H. Jakobs, 'Die Zeichen einer neuen Klasse, zur Typologie der frühen Stadtsiegel', in *Bild und Geschichte*, ed. K. Krimm and H. John (Sigmaringen, 1997), pp. 125–78
Dunbabin, J., 'The Reception and Interpretation of Aristotle's *Politics*', in *The Cambridge History of Later Medieval Philosophy: From the Rediscovery of Aristotle to the Disintegration of Scholasticism 1100–1600*, ed. N. Kretzman, A. Kenny and J. Pinborg (Cambridge, 1982), pp. 723–37
—, 'Guido Vernani of Rimini's Commentary on Aristotle's *Politics*', *Traditio*, XLIV (1988), pp. 373–88

Dutton, P. E., ed., *The Glosae super Platonem of Bernard of Chartres* (Toronto, 1991)
—, '*Illustre civitatis et populi exemplum*: Plato's *Timaeus* and the Transmission of Calcidius to the End of the Twelfth Century of a Tripartite Scheme of Society', *Mediaeval Studies*, XLV (1983), pp. 79–119
Eco, U., *The Aesthetics of Thomas Aquinas*, trans. H. Bredin (Cambridge, MA, 1988)
—, *Art and Beauty in the Middle Ages* (London, 1986)
Eliade, M., *The Sacred and the Profane, the Nature of Religion*, trans. W. R. Trask (London, 1959)
Eliassen, F.-E., and G. A. Ersland, ed., *Power, Profit and Urban Land: Landownership in Medieval and Early Modern Northern European Towns* (Aldershot, 1996)
Ennen, E., *The Medieval Town*, trans. N. Fryde (Amsterdam, 1979)
Fernie, E. C., 'The Romanesque Church of Bury St Edmunds Abbey', in *Bury St Edmunds. Medieval Art, Architecture, Archaeology and Economy* (London, 1998), ed. A. Gransden, pp. 1–15.
Forhan, K., trans., 'Brunetto Latini, *The Book of Treasure*', in *Medieval Political Theory – A Reader: The Quest for the Body Politic, 1100–1400*, ed. C. J. Nederman and K. Langdon Forhan (London, 1993), pp. 71–96
—, 'Christine de Pizan, *The Book of the Body Politic*', in *Medieval Political Theory – A Reader: The Quest for the Body Politic, 1100–1400*, ed. C. J. Nederman and K. Langdon Forhan (London, 1993), pp. 230–47
Frayling, C., *Strange Landscape: A Journey Through the Middle Ages* (London, 1995)
Friedman, D., *Florentine New Towns: Urban Design in the Late Middle Ages* (Cambridge, MA, 1988)
—, 'Palaces and the Street in Late Medieval and Renaissance Italy', in *Urban Landscapes: International Perspectives*, ed. J.W.R. Whitehand and P. J. Larkham (London, 1992), pp. 69–113
Friedman, J. B., 'The Architect's Compass in Creation Miniatures of the Later Middle Ages', *Traditio*, XXX (1974), pp. 419–29
—, *Monstrous Races in Medieval Art and Thought* (Cambridge, MA, 1981)
Frugoni, C., *A Distant City: Images of Urban Experience in the Medieval World*, trans. W. McCuiag (Princeton, NJ, 1991)
—, *Una lontana città. Sentimenti e immagini nel Medioevo* (Turin, 1983)
García Fernández, J. L., 'La rica experiencia urbanística del medievo español', in *La Ciudad Hispanoamericana. El Sueño de un Orden*, exh. cat., Ministerio de Obras Públicas y Urbanismo, Madrid (1989)
Gauthiez, B., *Atlas Morphologique des Villes de Normandie* (Lyon, 1999)
—, 'The Planning of the Town of Bury St Edmunds: A Probable Norman Origin', in *Bury St Edmunds: Medieval Art, Architecture, Archaeology and Economy* (London, 1998), ed. A. Gransden, pp. 81–97
Geremek, B., *The Margins of Society in Late Medieval Paris*, trans. J. Birrell (Cambridge, 1987)
Gibson, M., 'The Study of the *Timaeus* in the Eleventh and Twelfth Centuries', *Pensiamento*, XXV (1969), pp. 183–94
Giles, K., and C. Dyer, ed., *Town and Country in the Middle Ages: Contrasts, Contacts and Interconnections, 1100–1500* (Leeds, 2005)
Gimpel, J., *The Cathedral Builders*, trans. T. Waugh (London, 1983)
Gousset, M-T., 'La représentation de la Jérusalem céleste à l'époque Carolingienne', *Cahiers Archéologiques*, XXIII (1974), pp. 47–60
Gransden, A., ed., *Bury St Edmunds: Medieval Art, Architecture, Archaeology and Economy* (London, 1998)
—, 'Realistic Observation inTwelfth-Century England', *Speculum*, XLVII (1972), pp. 29–51

Grant, E., *The Foundations of Modern Science in the Middle Ages: Their Religious, Institutional, and Intellectual Contexts* (Cambridge, 1996)
—, *Physical Science in the Middle Ages* (Cambridge, 1971)
—, *Planets, Stars, and Orbs: The Medieval Cosmos, 1200–1687* (Cambridge, 1994)
Grau, F., ed., *Mentalitäten im Mittelalter: Methodische und inhaltliche Probleme* (Sigmaringen, 1987)
Gregory, T., 'The Platonic Inheritance', in *A History of Twelfth-Century Western Philosophy*, ed. P. Dronke (Cambridge, 1988), pp. 54–80
Greyerz, K. von, ed., *Religion and Society in Early Modern Europe 1500–1800* (London, 1984)
Grierson, P., *English Linear Measures: An Essay in Origins* (Reading, 1972)
—, 'The European Heritage', in *Ancient Cosmologies*, ed. C. Blacker and M. Loewe (London, 1975), pp. 225–57
Griffiths, R. A., ed., *Boroughs of Mediaeval Wales* (Cardiff, 1978)
Guidoni, E., *La Ville Européenne: Formation et signification du quatrième au onzième siècle* (Brussels, 1981)
Gunther, R. T., *Early Science in Oxford*, vol. I (Oxford, 1923)
Gurevich, A. J., *Categories of Medieval Culture*, trans. G. L. Campbell (London, 1985)
Gutkind, G., ed., *Urban Development in East-Central Europe: Poland, Czechoslovakia and Hungary* (London, 1972)
Hager, R., *The Zähringer New Towns*, exh. cat., Swiss Federal Institute of Technology, Zurich (1966)
Hall, R. A., 'The Five Boroughs of the Danelaw: A Review of Present Knowledge', *Anglo-Saxon England*, XVIII (1989), pp. 149–206
Halliwell, J., ed., *Rara Mathematica* (London, 1839)
Hamilton Bleakley, H., 'The Art of Ruling in Aquinas' *De Regimine Principum*', *History of Political Thought*, XX (1999), pp. 575–603.
Hammond, R., *The Philosophy of Alfarabi and its Influence on Medieval Thought* (New York, 1947)
Hanawalt, B., and K. L. Reyerson, ed., *City and Spectacle in Medieval Europe* (Minneapolis, MN, 1994)
—, 'Introduction', in *City and Spectacle in Medieval Europe*, ed. B. Hanawalt and K. L. Reyerson, pp. ix–xx.
Harley, J. B., and D. Woodward, ed., *The History of Cartography*, vol. I, *Cartography in Prehistoric, Ancient and Medieval Europe and the Mediterranean* (Chicago, IL, 1987)
Harty, K. J., ed., *The Chester Mystery Cycle: A Casebook* (London, 1993)
Harvey, J., *English Mediaeval Architects: A Biographical Dictionary Down to 1550* (Gloucester, 1984)
Harvey, P.D.A., 'Local and Regional Cartography in Medieval Europe', in *The History of Cartography*, vol. I, *Cartography in Prehistoric, Ancient and Medieval Europe and the Mediterranean*, ed. J. B. Harley and D. Woodward (Chicago, IL, 1987), pp. 464–501
Haskins, C. H., *Studies in the History of Medieval Science*, 2nd edn (Cambridge, MA, 1927)
Haslam, J., 'The Metrology of Anglo-Saxon Cricklade', *Medieval Archaeology*, XXX (1986), pp. 99–102
—, ed., *Anglo-Saxon Towns in Southern England* (Chichester, 1984)
Haverkamp, A., 'Heilige Städte im hohen Mittelater', in *Mentalitäten im Mittelalter*, ed. F. Grau (Sigmaringen, 1987), pp. 119–56
Hawkes, S. C., D. Brown and J. Campbell, ed., *Anglo-Saxon Studies in Archaeology and History*, vol. I (Oxford, 1979)

Heath, T. L., *History of Greek Mathematics*, vol. I (Oxford, 1921)
Higgins, A., 'Streets and Markets', in *A New History of Early English Drama*, ed. J. D. Cox and D. S. Kastan (New York, 1997), pp. 77–92
Hill, D., 'The Burghal Hidage: The Establishment of a Text', *Medieval Archaeology*, XIII (1969), pp. 84–92
Hillaby, J., 'The Norman New Town of Hereford: Its Street Pattern and European Context', *Transactions of the Woolhope Naturalists' Field Club*, XLIV (1982), pp. 181–95
Hillenbrand, R., "Anjar and Early Medieval Urbanism', in *The Idea and Ideal of the Town between Late Antiquity and the Early Middle Ages*, ed. G. P. Brogiolo and B. Ward-Perkins (Leiden, 1999), pp. 59–98
Hilton, R. H., *English and French Towns in Feudal Society: A Comparative Study* (Cambridge, 1992)
Hiscock, N., *The Symbol at your Door: Number and Geometry in the Religious Architecture of the Greek and Latin Middle Ages* (Aldershot, 2008)
—, *The Wise Master Builder: Platonic Geometry in Plans of Medieval Abbeys and Cathedrals* (Aldershot, 2000)
Hodges, R., *Towns and Trade in the Age of Charlemagne* (London, 2000)
Holmes, U. T., *Daily Living in the Twelfth Century: Based on the Observations of Alexander Neckam in London and Paris* (London, 1952)
Homann, F. A., trans., *Practical Geometry [Practica Geometriae] Attributed to Hugh of St Victor* (Milwaukee, WI, 1991)
Hopper, V. F., *Medieval Number Symbolism: Its Sources, Meaning, and Influence on Thought and Expression* (New York, 1969)
Horn, W., and E. Born, ed., *The Plan of St Gall: a Study of the Architecture and Economy of Life in a Paradigmatic Carolingian Monastery*, 3 vols (Berkeley, CA, 1979)
Hoy, J. F., 'On the Relationship of the Corpus Christi Plays to the Corpus Christi Procession at York', *Modern Philology*, LVVI (1973), pp. 166–8
Hughes, K., *The Church in Early Irish Society* (London, 1966)
Humpert, K., and M. Schenk, *Entdeckung der mittelalterlichen Stadtplanung: Das Ende vom Mythos der 'gewachsenen Stadt'* (Stuttgart, 2001)
Hyde, J. K., 'Medieval Descriptions of Cities', *Bulletin of the John Rylands Library*, XLVIII (1966), pp. 308–40
Jack, R. I., 'Ruthin', in *Boroughs of Mediaeval Wales*, ed. R. A. Griffiths (Cardiff, 1978), pp. 244–61
James, M., 'Ritual, Drama and Social Body in the Late Medieval English Town', *Past and Present*, XCVIII (1983), pp. 3–29
Jeauneau, É., ed., *Guillaume de Conches, Glosae super Platonem* (Paris, 1975)
Johnston, A. F., 'The Guild of Corpus Christi and the Procession of Corpus Christi in York', *Mediaeval Studies*, XXXVIII (1976), pp. 372–84
Kagan, R., 'A World Without Walls: City and Town in Colonial Spanish America', in *City Walls: The Urban Enceinte in Global Perspective*, ed. J. D. Tracy (Cambridge, 2000), pp. 117–52
Kalinowski, W., 'City Development in Poland', in *Urban Development in East-Central Europe: Poland, Czechoslovakia and Hungary*, ed. G. Gutkind (London, 1972), pp. 26–51
Kantorowicz, E., *The King's Two Bodies: A Study in Mediaeval Political Theology* (Princeton, NJ, 1957)
Karhl, S. J., 'The Staging of Medieval English Plays', in *The Theatre of Medieval Europe: New Research in Early Drama*, ed. E. Simon (Cambridge, 1991), pp. 130–48
—, *Traditions of Medieval English Drama* (London, 1974)

Katz, I., 'Les représentations de villes dans l'art Chrétien avant l'an mil', *Information de l'histoire de l'art*, III (1964), pp.

Keynes, S., and M. Lapidge, ed. and trans., *Alfred the Great: Asser's 'Life of Alfred' and other Contemporary Sources* (London, 1983)

Kibre, P., *Studies in Medieval Science: Alchemy, Astrology, Mathematics and Medicine* (London, 1984)

Kiely, E. R., *Surveying Instruments: Their History and Classroom Use* (New York, 1947)

Kimble, G.H.T., *Geography in the Middle Ages* (London, 1938)

Kitzinger, E., 'World Map and Fortune's Wheel: A Medieval Mosaic Floor in Turin', *Proceedings of the American Philosophical Society*, LXVII (1973), pp. 344–73

Klein, G. P., 'The Topography of Symbol: Between Late Antique and Modern Jewish Understanding of Cities', *Zeitschrift für Religions- und Geistesgeschichte*, LVIII (2006), pp. 16–28

Klibansky, R., *The Continuity of the Platonic Tradition*, new edn (New York, 1982)

Knight, A. E., ed., *The Stage as Mirror: Civic Theatre in Late Medieval Europe* (Cambridge, 1997)

Kohl, B. G., *Padua under the Carrara, 1318–1405* (London, 1998)

Kolve, V. A., *The Play Called Corpus Christi* (Stanford, CA, 1965)

Koter, M., and M. Kulesza, 'The Plans of Medieval Polish Towns', *Urban Morphology*, III (1999), pp. 63–78

Krush, B., and W. Levison, ed., *Gregory of Tours, Historiarum libri X*, in *Gregorii episcopi turnonsis Historiarum libri X, Germaniae Historica, Scriptores rerum Merovingicarum*, i.1 (Hanover, 1951)

Kühnel, B., *From the Earthly to the Heavenly Jerusalem: Representations of the Holy City in Christian Art of the First Millennium* (Rome, 1987)

—, 'Geography and Geometry of Jerusalem', in *City of the Great King: Jerusalem from David to the Present*, ed. N. Rosovsky (Cambridge, MA, 1996), pp. 288–332

—, ed., *The Real and Ideal Jerusalem in Jewish, Christian and Islamic Art* (Jerusalem, 1998)

—, 'The Use and Abuse of Jerusalem', in *The Real and Ideal Jerusalem in Jewish, Christian and Islamic Art* (Jerusalem, 1998), ed. B. Kühnel, pp. xix–xxxviii

Ladner, G. B., 'Medieval and Modern Understanding of Symbolism: A Comparison', *Speculum*, LIV (1979), pp. 223–56

—, 'The Symbolism of the Biblical Corner Stone in the Mediaeval West', *Mediaeval Studies*, IV (1942), pp. 43–60

Lanman, J. T., 'The Religious Symbolism of the T in T-O maps', *Cartographica*, XVIII (1981), pp. 18–22

Lauret, A., R. Malebranche and G. Séraphin, *Bastides, Villes Nouvelles du Moyen Age* (Toulouse, 1988)

Laveden, P., *Représentation des villes dans l'art du Moyen Age* (Paris, 1954)

Lawlor, H. J., *Chapters in the Book of Mulling* (Edinburgh, 1897)

Leach, A. F., ed., *Beverley Town Documents*, Selden Society, XIV (London, 1900)

—, *Report on the Manuscripts of the Corporation of Beverley* (London, 1900)

Lee, D., trans., *Plato, Timaeus and Critias* (London, 1977)

Lesser, G., *Gothic Cathedrals and Sacred Geometry* (London, 1957), 2 vols

Lethaby, W. R., *Mediæval Art: From the Peace of the Church to the Eve of the Renaissance 312–1350* (London, 1912)

Levy, R. I., *Mesocosm: Hinduism and the Organisation of a Traditional Newar City in Nepal* (Berkeley, CA, 1990)

Levy-Rubin, M., and R. Rubin, 'The Image of the Holy City in Maps and Mapping', in *City of the Great King: Jerusalem from David to the Present*, ed. N. Rosovsky (Cambridge, MA, 1996), pp. 352–79
Lewis, C. P., and A. T. Thacker, ed., *A History of the County of Chester*, vol. v/1, *The City of Chester: General History and Topography* (London, 2003)
Lewis, M.J.T., *Surveying Instruments of Greece and Rome* (Cambridge, 2001)
Lilley, K. D., 'Cities of God? Medieval Urban Forms and their Christian Symbolism', *Transactions of the Institute of British Geographers*, new series, XXIX (2004), pp. 296–313.
—, 'Mapping Cosmopolis: Moral Topographies of the Medieval City', *Environment and Planning d: Society and Space*, XXII (2004), pp. 681–98
—, 'Mapping the Medieval City: Plan Analysis and Urban History', *Urban History*, XXVII (2000), pp. 5–30
—, 'Modern Visions of the Medieval City: Competing Conceptions of Urbanism in European Civic Design', *Environment and Planning b: planning and design*, XXVI (1999), pp. 427–46
—, '*Non urbe, non vico, non castris*: Territorial Control and the Colonization and Urbanization of Medieval Wales and Ireland under Anglo-Norman Lordship', *Journal of Historical Geography*, XXVI (2000), pp. 517–31
—, 'Reading the Medieval Urban Landscape: Approaches to the Morphological Study of Landscape', in *Reading the Landscape*, ed. D. Munro (Edinburgh, 2002), pp. 70–82
—, 'Taking Measures Across the Medieval Landscape: Aspects of Urban Design before the Renaissance', *Urban Morphology*, II (1998), pp. 82–92
—, 'Urban Landscapes and their Design: Creating Town from Country in the Middle Ages', in *Town and Country in the Middle Ages*, ed. K. Giles and C. Dyer (Leeds, 2005), pp. 229–49
—, *Urban Life in the Middle Ages, 1000–1450* (London, 2002)
—, 'Urban Planning and the Design of New Towns in the Middle Ages: The Earls of Devon and their "New Towns"', *Planning Perspectives*, XVI (2001), pp. 1–24
Limor, O., 'The Place of the End of Days: Eschatological Geography in Jerusalem', in *The Real and Ideal Jerusalem in Jewish, Christian and Islamic Art*, ed. B. Kühnel (Jerusalem, 1998), pp. 13–22
Lindberg, D. C., ed., *Science in the Middle Ages* (Chicago, 1978)
—, 'The Transmission of Greek and Arabic Learning to the West', in *Science in the Middle Ages*, ed. D. C. Lindberg (Chicago, 1978), pp. 52–90
Lindenbaum, S., 'Ceremony and Oligarchy: The London Midsummer Watch', in *City and Spectacle in Medieval Europe*, ed. B. Hanawalt and K. Reyerson (Minneapolis, MN, 1994), pp. 171–88
Lloyd, G.E.R., 'Greek Cosmologies', in *Ancient Cosmologies*, ed. C. Blacker and M. Loewe (London, 1975), pp. 206–7
Lobel, M. D., ed., *The Atlas of Historic Towns: Bristol, Cambridge, Coventry, Norwich* (London, 1975)
—, 'Notes on the History of Oxford', *Oxoniensia*, III (1938), pp. 83–102
Mâle, E., *Religious Art in France: The Thirteenth Century* (Princeton, NJ, 1984)
Marenbon, J., ed., *Routledge History of Philosophy*, vol. III, *Medieval Philosophy* (London, 1988)
Martin, C., 'Some Medieval Commentaries on Aristotle's *Politics*', *History*, XXXVI (1951), pp. 29–44
McRee, B. R., 'Unity or Division? The Social Meaning of Guild Ceremony in Urban Communities', in *City and Spectacle in Medieval Europe*, ed. B. Hanawalt and K. Reyerson (Minneapolis, MN, 1994), pp. 189–207

Meckseper, C., *Kleine Kunstgeschichte der Deutschen Stadt im Mittelalter* (Darmstadt, 1982)
Mendthal, H., ed., *'Geometria culmensis', ein agronomischer Traktat aus der Zeit des Hochmeisters Conrad von Jungingen (1393–1407)* (Leipzig, 1886)
Meredith, P., and J. E. Tailby, ed., *The Staging of Religious Drama in Europe in the Later Middle Ages: Texts and Documents in Translation* (Kalamazoo, MI, 1983)
Mill, A. J., 'The Stations of the York Corpus Christi Play', *Yorkshire Archaeological Journal*, XXXVII (1951), pp. 492–502
Miller, W., ed. *Cicero, De Officiis* (London, 1913)
Morris, A.E.J., *History of Urban Form: Before the Industrial Revolutions* (Harlow, 1994)
Morrissey, L. J., 'English Pageant-wagons', *Eighteenth-Century Studies*, IX (1976), pp. 353–74
Muckle, J. T., 'Greek Works Translated Directly into Latin before 1350', *Mediaeval Studies*, IV (1942), pp. 33–42, 102–114
Muir, L. R., *The Biblical Drama of Medieval Europe* (Cambridge, 1995)
Munby, J., 'Saxon Chichester and its Predecessors', in *Anglo-Saxon Towns in Southern England*, ed. J. Haslam (Chichester, 1984), pp. 315–30
Mundy, J. H., 'In Praise of Italy: The Italian Republics', *Speculum*, LXIV (1989), pp. 815–34
Munro, D., ed., *Reading the Landscape* (Edinburgh, 2002)
Nagler, A. M., *The Medieval Religious Stage: Shapes and Phantoms* (London, 1976)
Nas, P.M.J., ed., *Urban Symbolism*, Leiden Studies in Human Society (Leiden, 1993)
Nederman, C. J., ed. and trans., *John of Salisbury, 'Policraticus'* (Cambridge, 1990)
—, trans., 'Marsiglio of Padua, *The Defender of the Peace*', in *Medieval Political Theory – A Reader: The Quest for the Body Politic, 1100–1400*, ed. C. J. Nederman and K. Langdon (London, 1993), pp. 173–99
—, and Langdon, K., eds., *Medieval Political Theory – A Reader: The Quest for the Body Politic, 1100–1400* (London, 1993)
—, 'Nature, Sin and the Origins of Society: The Ciceronian Tradition in Medieval Political Thought', *Journal of the History of Ideas*, XLIX (1988), pp. 3–26
Nelson, A. H., *The Medieval English Stage: Corpus Christi Pageants and Plays* (Chicago, IL, 1974)
Nicholas, D., *The Growth of the Medieval City* (London, 1997).
Nijsten, G., 'Feasts and Public Spectacle: Late Medieval Drama and Performance in the Low Countries', in *The Stage as Mirror: Civic Theatre in Late Medieval Europe*, ed. A. E. Knight (Cambridge, 1997), pp. 107–44
Nitz, H.-J., 'Planned Temple Towns and Brahmin Villages as Spatial Expressions of the Ritual Politics of Medieval Kingdoms in South India', in *Ideology and Landscape in Historical Perspective*, ed. A. Baker and G. Biger (Cambridge, 1992), pp. 107–24
Obrist, B., 'Wind Diagrams and Medieval Cosmology', *Speculum*, LXXII (1997), pp. 58–65
Ottaway, P., *Archaeology in British Towns: From the Emperor Claudius to the Black Death* (London, 1992)
Ousterhout, R., 'Flexible Geography and Transportable Topography', in *From the Earthly to the Heavenly Jerusalem: Representations of the Holy City in Christian Art of the First Millennium*, ed. B. Kühnel (Rome, 1987), pp. 393–404.
Owen, D. H., 'Denbigh', in *Boroughs of Mediaeval Wales*, ed. R. A. Griffiths (Cardiff, 1978), pp. 164–87
Palliser, D. M., ed., *The Cambridge Urban History of Britain*, vol. I: *600–1540*

(Cambridge, 2000)
—, *Chester: Contemporary Descriptions by Residents and Visitors* (Chester, 1980)
Panofsky, E., *Gothic Architecture and Scholasticism* (Latrobe, PA, 1951)
Passini, J., *El Camino de Santiago. Itinerarios y núcleos de poblacióu* (Madrid, 1993)
Pawlowski, K., 'Villes et villages circulaires du Languedoc', *Annales du Midi*, IX (1987), pp. 407–28
Payne, A. F., *King Alfred and Boethius: An Analysis of the OE Version of the 'Consolation of Philosophy'* (Madison, WI, 1968)
Pedrick, G., *Borough Seals of the Gothic Period* (London, 1904)
Phythian Adams, C., 'Ceremony and the Citizen: The Communal Year at Coventry 1450–1550', in *Crisis and Order in English Towns*, ed. P. Clark and P. Slack (London, 1972), pp. 57–85
—, *Desolation of a City: Coventry and the Urban Crisis of the Late Middle Ages* (Cambridge, 1979)
Pirenne, H., *Medieval Cities: Their Origins and the Revival of Trade*, trans. F. D. Halsey (Princeton, NJ, 1925)
Posèq, A.W.G., 'The "New Jerusalem", the Star of Zion and the Mandala', in *The Real and Ideal Jerusalem in Jewish, Christian and Islamic Art*, ed. B. Kühnel (Jerusalem, 1998), pp. 325–39
Potter, K. R., ed. and trans., *Gesta Stephani* (Oxford, 1976)
Powlick, L., 'The Staging of the Chester Cycle: An Alternate Theory', in *The Chester Mystery Cycle: A Casebook*, ed. K. J. Harty (London, 1993), pp. 199–229
Prestwich, M., *Edward I* (New Haven, CT, 1988)
Quirin, H., 'The Colonial Town as Seen in the Documents of East German Settlement', in *The Comparative History of Urban Origins in Non-Roman Europe*, ed. H. B. Clarke and A. Simms (Oxford, 1985), pp. 505–29.
Raleigh Radford, C. A., 'The Later Pre-Conquest Boroughs and their Defences', *Medieval Archaeology*, XIV (1970), pp. 83–103
Ralph, E., 'Bristol, *circa* 1480', in *Local Maps and Plans from Medieval England*, ed. R. A. Skelton and P.D.A. Harvey (Oxford, 1986), pp. 309–16
Randolph, A., '*Bastides* of South-west France', *Art Bulletin*, LXXVII (1995), pp. 290–307.
Rawcliffe, C., 'The Earthly and Spiritual Topography of Suburban Hospitals', in *Town and Country in the Middle Ages*, ed. K. Giles and C. Dyer (Leeds, 2005), pp. 251–74
Reuter, T., ed., *Alfred the Great: Papers from the Eleventh-Centenary Conferences* (Aldershot, 2003)
Reynolds, S., *An Introduction to the History of English Medieval Towns* (London, 1977)
—, *Kingdoms and Communities in Western Europe, 900–1300* (Oxford, 1984)
—, 'Medieval Urban History and the History of Political Thought', *Urban History Yearbook* (1982), pp. 14–23
Riley, H. T., ed. and trans., *Munimenta Gildhallæ Londoniensis; Liber Albus, Liber Custumarum, et Liber Horn*, Rolls Series, vol. II, pt. 2 (1860)
Ringbom, L. I., *Graltempel und Paradies* (Stockholm, 1951)
Rörig, F., *The Medieval Town*, trans. D. Bryant (London, 1967)
Rosenau, H., *The Ideal City* (London, 1983)
Rosovsky, N., ed., *City of the Great King: Jerusalem from David to the Present* (Cambridge, MA, 1996)
Rothwell, H., ed., *English Historical Documents 1189–1327* (London, 1975)
Rubin, M., *Corpus Christi: The Eucharist in Late Medieval Culture* (Cambridge, 1991)

Ryckaert, M., *Historische Stedenatlas van België. Brugge* (Brussels, 1991)
Rykwert, J., *The Idea of a Town: The Anthropology of Urban Form in Rome, Italy and the Ancient World* (Princeton, NJ, 1988)
Salmon, D., 'The Mediæval Latin Translations of Alfarabi's Works', *New Scholasticism*, XIII (1939), pp. 245–61
Salter, H. E., *Medieval Oxford* (Oxford, 1936)
Salvemini, C., ed., *Giovanni da Viterbo, Liber de Regimine Civitatum, in Biblioteca juridical medii aevi*, vol. III(Bologna, 1901)
Sandred, K. I., and B. Lingström, *The Place-Names of Norfolk*, part I (Irthingborough, 1989)
Schmitt, M., *Aachen, Deutscher Städtatlas 4* (Altenbeken, 1984)
Scholkmann, B., 'Zwischen Mythos und Befund. Eine kritische Bilanz zum Thema "Die vermessene Stadt" aus Sicht der archäologischen Stadtkernforschung', *Mitteilungen der Deutschen Gesellschaft für Archäologie des Mittelalters und der Neuzeit* (Paderborn, 2004), pp. 180–84
Scholz, R., ed., *Marsilius of Padua, Defensor pacis*, in *MGH*, *Fontes iuris Germanici antique in usum scholarum* (Hanover, 1932)
Schulz, J., 'Jacopo de' Barbari's View of Venice: Map Making, City Views and Moralized Geography before the Year 1500', *Art Bulletin*, LX (1978), pp. 425–74
Scribner, B., 'Cosmic Order and Daily Life: Sacred and Secular in Pre-industrial German Society', in *Religion and Society in Early Modern Europe 1500–1800*, ed. K. von Greyerz (London, 1984), pp. 17–32
Sedgefield, W. J., ed. and trans., *King Alfred's Old English Version of Boethius' 'De Consolatione Philosophiae'* (Oxford, 1899)
Sellers, M., ed., *York Memorandum*, Book 2 (Durham, 1912–15)
Seyer, S., ed., *Charters and Letters Patent Granted by the Kings and Queens of England to the Town and City of Bristol* (Bristol, 1812)
Shelby, L. R., 'The Education of Medieval English Master Masons', *Mediaeval Studies*, XXXII (1970), pp. 1–26
—, 'The Geometrical Knowledge of the Mediaeval Master Masons', *Speculum*, XLVII (1972), pp. 395–421
—, 'Medieval Masons' Tools II. Compass and Square', *Technology and Culture*, VI (1965), pp. 236–48
Sheridan, J. J., trans., *Alan of Lille: The Plaint of Nature* (Toronto, 1980)
Shopkow, L., ed. and trans., *Lambert of Ardres, The History of the Counts of Guines and Lords of Ardres* (Philadelphia, PA, 2001)
Sigmund, P. E., ed. and trans., *St Thomas Aquinas on Politics and Ethics* (New York, 1988)
—, *Nicholas of Cusa and Medieval Political Thought* (Cambridge, MA, 1963)
Simon, E., ed., *The Theatre of Medieval Europe. New Research in Early Drama* (Cambridge, 1991)
Sinclair, T. A., ed. and trans., *Aristotle, The Politics* (London, 1962)
Skelton, R. A., and P.D.A. Harvey, ed., *Local Maps and Plans from Medieval England* (Oxford, 1986)
Skinner, G. W., ed., *The City in Late Imperial China* (Stanford, CA, 1977)
Slater, T. R., ed., *The Built Form of Western Cities* (Leicester, 1990)
—, 'English Medieval New Towns with Composite Town Plans: Evidence from the Midlands', in *The Built Form of Western Cities*, ed. T. R. Slater (Leicester, 1990), pp. 60–82
—, 'Ideal and Reality in English Episcopal Medieval Town Planning', *Transactions of the Institute of British Geographers*, new series, XII (1987), pp. 191–203

—, 'Medieval Town-founding on the Estates of the Benedictine Order in England', in *Power, Profit and Urban Land: Landownership in Medieval and Early Modern Northern European Towns*, ed. F.-E. Eliassen and G. A. Ersland (Aldershot, 1996), pp. 70–92

Smet, de, A., 'De l'utilité de recueillir les mentions d'arpenteurs cités dans les documents d'archives du moyen âge', *Annales du Congrès-Archéologique et Historique de Tournai*, III (1949), pp. 782–95

Smith, A. H., *Place-Names of Gloucestershire* (Cambridge, 1964)

Smith, L. T., *The Maire of Bristowe is Kalendar*, Camden Society, new series, V (1872)

—, ed., *York Plays: The Plays Performed by the Crafts or Mysteries of York on the Day of Corpus Christi* (Oxford, 1885)

Snape, M. G., 'Durham 1439 x circa 1442', in *Local Maps and Plans from Medieval England*, ed. R. A. Skelton and P.D.A. Harvey (Oxford, 1986), p. 189–94

Southall, A., 'The Circle and the Square: Symbolic Form and Process in the City', in *Urban Symbolism*, ed. P.M.J. Nas, Leiden Studies in Human Society (Leiden, 1993), pp. 378–93

Southern, R. W., *Platonism, Scholastic Method and the School of Chartres* (Reading, 1978)

—, *Scholastic Humanism and the Unification of Europe* (Oxford, 1995)

Spencer, M., *Corpus Christi Pageants in England* (New York, 1911)

Stahl, W. H., trans., *Macrobius' Commentary on the Dream of Scipio* (New York, 1952)

—, *Roman Science: Origins, Development and Influence to the Later Middle Ages* (Madison, WI, 1962)

Staines, D., 'The English Mystery Cycles', in *The Theatre of Medieval Europe*, ed. E. Simon (Cambridge, 1991), pp. 80–96

Steele, R., ed., *Communia mathematica fratris Rogeri [Roger Bacon]* (Oxford, 1940)

Steinhardt, N., *Chinese Imperial City Planning* (Honolulu, 1992)

Stephens, W. B., ed., *A History of the County of Warwick*, vol. VIII: *The City of Coventry and Borough of Warwick* (London, 1969)

Stevenson, W. H., ed., *Asser's Life of Alfred* (Oxford, 1904)

Stock, B., *Myth and Science in the Twelfth Century: A Study of Bernard Silvester* (Princeton, NJ, 1972)

Swan, L., 'Monastic Proto-towns in Early Medieval Ireland: The Evidence of Aerial Photography, Plan Analysis and Survey', in *The Comparative History of Urban Origins in Non-Roman Europe*, ed. H. B. Clarke and A. Simms (Oxford, 1985), pp. 77–102

Tannery, P., 'Géométrie', *Revue de Synthèse Historique*, II (1901), pp. 283–99

—, 'La géométrie au XIe siècle', *Revue Générale Internationale Scientifique, Litteraire et Artistique*, XV (1897), pp. 343–57

—, 'Le Traité du quadrant de maître Robert Anglès', *Notices et extraits des manuscrits de la Bibliothèque Nationale*, XXXV (1897), pp. 561–640

Tatton-Brown, T., 'The Topography of Anglo-Saxon London', *Antiquity*, LX (1986), pp. 21–8

Taylor, J., trans., *The Didascalicon of Hugh of St Victor: A Medieval Guide to the Arts* (New York, 1961)

—, 'The Dramatic Structure of the Middle English Corpus Christi, or Cycle, Plays', in *Medieval English Drama: Essays Critical and Contextual*, ed. J. Taylor and A. H. Nelson (London, 1972), pp. 148–56

—, and A. H. Nelson, ed., *Medieval English Drama: Essays Critical and Contextual* (London, 1972)

Taylor, M. V., ed., *Liber Luciani de Laude Cestrie*, Lancashire and Cheshire Record Society, LXIV (1912)
Terry, M. S., trans., *The Sibylline Oracles, Translated from the Greek into English Blank Verse* (New York, 1899)
Thorndike, L., ed. and trans., *The 'Sphere' of Sacrobosco and its Commentators* (Chicago, 1949)
Thulin, C., ed., *Corpus Agrimensorum Romanorum*, vol. I, fasc. I, *Opuscula agrimensorum veterum* (Leipzig, 1913)
Tingey, J. C., ed., *The Records of the City of Norwich*, vol. II (Norwich, 1910)
Trabut-Cussac, J. P., 'Date, fondation et identification de la *bastide* de Baa', *Revue Historique de Bordeaux*, X (1961), pp. 133–44
Tracy, J. D., ed., *City Walls: The Urban Enceinte in Global Perspective* (Cambridge, 2000)
Travis, P. W., 'The Social Body of the Dramatic Christ in Medieval England', *Early English Drama. Acta*, XIII, pp. 17–36
Twycross, M., ed., *Festive Drama* (Cambridge, 1996)
—, 'Some Approaches to Dramatic Festivity, Especially Processions', in *Festive Drama*, ed. M. Twycross (Cambridge, 1996), pp. 1–33
Tydeman, W., *The Theatre in the Middle Ages* (Cambridge, 1978)
Ullman, B. L., 'Geometry in the Mediaeval Quadrivium', in *Studi de Bibliografia e di Storia in Onore di Tammaro de Marinis*, ed. T. de Marinis (Verona, 1964), pp. 263–85
Verhulst, A., *The Rise of Cities in North-West Europe* (Cambridge, 1999)
Victor, S. K., ed. and trans., *Practical Geometry in the High Middle Ages: 'Artis Cuiuslibet Consummatio' and the 'Pratike de Geometrie'* (Philadelphia, PA, 1979)
Vince, A., *Saxon London: An Archaeological Investigation* (London, 1990)
Wagner, D. L., ed., *The Seven Liberal Arts in the Middle Ages* (Bloomington, IN, 1983)
Waszink, J. H., ed., *Timaeus a Calcidio translatus commentarioque instructus*, 2nd edn (London, 1975)
Watts, V., ed. and trans., *Boethius, The Consolation of Philosophy* (London, 1999)
Weiss, D. H., '*Hec est Domus Domini firmiter edificata*: The Image of the Temple in Crusader Art', in *The Real and Ideal Jerusalem in Jewish, Christian and Islamic Art*, ed. B. Kühnel (Jerusalem, 1998), pp. 210–17
Westra, H. J., ed., *From Athens to Chartres: Neoplatonism and Medieval Thought: Studies in Honour of Edouard Jeauneau* (Leiden, 1992)
Wetherbee, W., 'Philosophy, Cosmology and the Twelfth-Century Renaissance', in *A History of Twelfth-Century Western Philosophy*, ed. P. Dronke (Cambridge, 1988), pp. 21–53
Wheatley, P., *City as Symbol* (London, 1969)
—, *The Pivot of the Four Quarters: A Preliminary Enquiry into the Origins and Character of the Ancient Chinese City* (Edinburgh, 1971)
White, E., 'Places for Hearing the Corpus Christi Play in York', *English Medieval Theatre*, IX (1987), pp. 23–63
Whitehand, J.W.R., ed., *The Urban Landscape: Historical Development and Management: Papers by M.R.G. Conzen* (London, 1981)
—, 'Background to the Urban Morphogenetic Tradition', in *The Urban Landscape: Historical Development and Management: Papers by M.R.G. Conzen*, ed. J.W.R. Whitehand (London, 1981), pp. 1–24
—, and Larkham, P. J., ed., *Urban Landscapes: International Perspectives* (London, 1992)
Wickham, G., *Medieval Theatre* (Cambridge, 1987)

Wilks, M., ed., *The World of John of Salisbury* (Oxford, 1984)
Williams, A., and G. H. Martin, ed. and trans., *Domesday Book, a Complete Translation* (London, 2002)
Williams, G. A., *Medieval London from Commune to Capital* (London, 1963)
Woodward, D., 'Medieval *mappaemundi*', in *The History of Cartography*, vol. I: *Cartography in Prehistoric, Ancient and Medieval Europe and the Mediterranean*, ed. J. B. Harley and D. Woodward (Chicago, IL, 1987), pp. 286–370
Wright, A. F., 'The Cosmology of the Chinese City', in *The City in Late Imperial China*, ed. G. W. Skinner (Stanford, CA, 1977), pp. 33–73
Zagrodski, T., 'Les plans de villes créées au moyen âge en Pologne et la tradition des règles gromaticales', *Revue des Archéologiques et Historiens d'Art de Louvain*, IX (1976), pp. 105–18
Zaitsev, E. A., 'The Meaning of Early Medieval Geometry, from Euclid and Surveyors' Manuals to Christian Philosophy', *Isis*, XC (1999), pp. 522–53
Zientara, B., 'Socio-economic and Spatial Transformation of Polish Towns during the Period of Location', *Acta Poloniae Historica*, XXXIV (1976), pp. 57–83
Zika, C., 'Hosts, Processions and Pilgrimages: Controlling the Sacred in Fifteenth-century Germany', *Past and Present*, CXVIII (1988), pp. 25–64
Ziller, S., C. Nederman and K. Forhan, ed. and trans., 'Thomas of Aquinas, *On Kingship*, *Summary of Theology* and *Commentary on Aristotle's Politics*', in *Medieval Political Theory – A Reader: The Quest for the Body Politic, 1100–1400*, ed. C. J. Nederman and K. Langdon Forhan (London, 1993), pp. 97–148

ACKNOWLEDGEMENTS

In the course of researching and writing this book I have incurred numerous debts. It began to take shape at Royal Holloway (University of London) thanks to a British Academy Post-Doctoral Fellowship, and has been written while in post at Queen's University Belfast. Parts of it, and the ideas stated within it, have been variously presented at conferences and seminars at universities in North America and in Europe. In particular, criticisms gained from presenting this work to distinguished audiences at the 'Dimensionen Symbolischer Sinnstiftung' colloquium held at the Katholisch-Theologische Fakultät of the University of Münster, and the 'Religion and the State' colloquium held at the Centre for History and Economics at the University of Cambridge, were especially helpful, and I am grateful to Ulrich Fischer and Miri Rubin for their kind invitations to speak at these seminars and for their comments. The book has also gained from discussions arising with seminar participants at the Universities of Birmingham, Dublin, Exeter, London, Nottingham, Oxford and York. For their honest and informed views on aspects of this book I am especially indebted to Wim Boerfijn, Tom Boogaart, Brian Campbell, Bruce Campbell, Denis Cosgrove, Christopher Dyer, Christian Frost, Richard Holt, David Livingstone, Rev. John Medhurst, David Palliser, Anngret Simms and Canon Terry Slater. I am grateful too for editorial comments and advice received from Michael Leaman, and also to Gill Alexander, who redrew many of the illustrations that appear in this book. The book could not have been written without the support of my wife, Catherine, and it is dedicated to my mother, who died before I could complete it.

PHOTO ACKNOWLEDGEMENTS

The author and publishers wish to express their thanks to the below sources of illustrative material and/or permission to reproduce it.

Gill Alexander: 1, 5, 6, 10, 11, 12, 14, 15, 17, 18, 19, 20, 21, 22, 23, 25, 27, 28, 30, 31, 35, 54, 56, 57, 58, 59, 60, 63, 64; redrawn from photograph by Gill Alexander: 34; Archivio di Stato, Siena: 49 (Capitoli 3, fols. 25v–26); Bayerische Staatsbibliothek, Munich: 43 (clm. 16128, fol. 27r), 45 (clm., 210, fol. 132v), 46 (clm., 16128, fol. 16r); Bibliotheque Municipale, Valenciennes: 38 (MS 99, fol. 38r); Bibliothèque Nationale de France, Paris (French Collection, MS 19093): 33; Bristol Record Office (photo © Bristol Record Office): 42 (MS 04720, fol. 5v); British Library, London (photos © British Library Board, all rights reserved): 2 (MS Add. 32343, fol. 15), 3 (Harley 658, fol. 39v), 9 (Cotton Nero D I, fol. 23v), 36 (Add. MS 28681, fol. 9); British Museum, London (photos © The Trustees of the British Museum): 52 (MLA 1972, 1–4.1), 53 (MLA 1909, 0617.1); Dioezesan- und Dombibliothek, Köln: 8 (Cod. 83-II, fol. 84r); reproduced courtesy Fondation de l'Oeuvre Notre-Dame, Strasbourg: 32; Herzog August Bibliothek Wolfenbüttel: 40 (Cod. Guelf. 1 Gud. Lat. 2°, 43v), 50 (Cod. Guelf. 36.23 Aug. 2°, 49v), 55 (Cod. Guelf. 36.23 Aug. 2°, 54r), 51 (Cod. Guelf. 36.23 Aug. 2°, 42r and 42v); Koninklijke Bibliotheek, The Hague: 41 (MS 76, fol. 5); photos Keith Lilley: 13, 16, 26, 61, 65; Österreichische National Bibliothek, Vienna (photo © Austrian National Library Vienna, Picture Archive): 47 (Codex Vindobonensis 2554); from Gale Pedrick, *Borough Seals of the Gothic Period* (London, 1904): 4; photos Terry Slater: 24, 29, 62; Stadtbibliothek, Trier: 7 (MS 1084/115 4°, fol. 99r), 37 (MS 31, fol. 69r), Trinity College, Cambridge (photo © Master and Fellows of Trinity College): 39 (MS R 16 2); Trinity College, Dublin, Library (photo © The Board of Trinity College): 48 (MS 177, fol. 59v).

INDEX

Towns and cities are listed under individual countries and regions.

'Adam', letters signifying four cardinal points 132
Adelard of Bath 114–15, 116
aesthetics and geometrical built forms 60–73
  Florentine 'sine geometry' 70–71
  geometric schema, aesthetic significance of 34–5
  geometry and beauty 62
  making the 'invisible' visible 71–3
  numerical ratios *30*, 61, 65, 66, 67–9, 72, 90
  orthogonal forms *25*, 62–5, 66–7, 68
  proportioned forms *15*, *17*, *22*, *27*, *28*, *29*, 49–50, 61–2, 63, 65–72, 82, 89, 90, 93
  root 2 proportions *30*, 68–9, 90
  rotating square method 68, 69–70
  straighter streets *15*, *25*, 63, 72–3
  symbolic purpose of 64–5, 68, 72–3
agrimensores, *Corpus Agrimensorum Romanorum* (Roman surveying manuals) *50*, *51*, *54*, *55*, 86, 110, 111, 116–21, 122, 123, 186
Alan of Lille 84, 142, 186
  Plaint of Nature 9–10, 11, 33, 36–7, 79
Albert the Great, *De Pulchro et Bono* 71
Alfred the Great 21, 79
  and *burhs see burhs* of Anglo-Saxon England
altimetry 40, 95, 96, 118, 123, 124
Aquinas, Thomas 8, 89, 186
  Commentary on the Politics 79, 139–40
  De Regimine Principum 79–80, 81, 82, 139, 141, 144
  Summa 71
Arcerianus manuscript 121
architects
  architectural metaphors and Neoplatonism 39, 80–81, 126
  God as architect *9*, 36, 37–40, *47*, *48*, 77, 80–81, 85, 108, 109, 126–7
  urban design and geometrical knowledge *9*, *32*, *33*, *48*, 86–90, 109
Aristotle 29, 35, 80, 137, 142, 186
  Metaphysics 81
  Physics 8
  Politics 79, 81, 138, 140–41
Artis Cuiuslibet Consummatio 88, 96, 113, 121, 123, 124, 125
astrolabe, use of 40, *53*, 112, 124–5
Augustine 26, 34, 81, 142
  City of God 27, 37, 38, 185
  Enarrationes in Psalmos 132
  Soliloquies 45
Austria, Vienna, circular shape 21

Bacon, Roger 39, 125, 127
  Communia Mathematica 87
bastides, France
  charter granting 93
  layout variations *15*, *16*, *22*, 48, 57–9, 63, 67, 68–9, 90, 91–2, 93
  place, positioning of *17*, 48–50
  proportional design layout *15*, *17*, *22*, *30*, 49–50, 67, 68
  rectilinear layout of *13*, *14*, *18*, *28*, 46–8, 50–51, 52, 63
  see also France
Bede, Venerable 31
  De Natura Rerum 34, *45*, 106
Belgium *see under* Flanders
Bernard of Chartres 29
Bernard of Silvester 33
body metaphors, and moral topographies 135, 142–3, 144
Boethius 25, 28, 34, 61, 72, 81, 86, 117, 142

Consolation of Philosophy 114
Consolations 45, 79
translation of Euclid 114
Breteuil, Law of *56*, *57*, 146–9, 150
burhs of Anglo-Saxon England 121
irregularities in 62–3
and Jerusalem 45
models for 44
and placed identities 57, 150–51
and Roman *castra* 44
square form *10*, *11*, 42–5, 51, 52
unit of measure 43, 121
see also Alfred the Great; England

Calcidius 25, 28, 72, 78, 86, 137, 145, 153
Commentary on Plato's Timaeus 6, 8–11, 12, 29, 33, 142
Cambiuzzi, Bachino 89
Cambrai manuscript 16, *37*, 98
Carolingian period 15–16, 17, 35, 44, 117, 121
Cassiodorus 28, 114
Charlemagne 44
Christian imaginings of medieval world 28–35
Christian kingship and sovereign rule 78–9
Christian thinking and Neoplatonism 8, 9, 81, 114
Christine de Pizan, *Book of the Body Politic* 142
Cicero 137, 138, 139, 140
De Officiis 141–2
circular built forms
circle and square, combining 20–21, *21*, *22*, *23*, *24*, *42*, *49*, 56–60, 63, 82, 89, 92, 103, 110, 126
circular geometries and their symbolism *20*, 53–6
Flanders, Bruges, circular-shaped outer defensive circuit 53
France *21*, 56
Ireland, Armagh, circular monastic town *20*, 54, 55–6
Italy, circle with square form *49*, 56, 57, 59, 82, 89, 110, 126
Jerusalem *2*, *3*, 15, 16–19, 20, 22–3, 27, 37, *36*, *38*, *41*, *44*, 54, 56, 97, 99, 102, 105
Jerusalem, circular and square shapes, reconciling 7, *8*, 18, 20, 21, 25, 32, 35, 56
scarcity of, in twelfth and thirteenth centuries 53–6
see also square built forms
Clarembald of Arras 38
Classical legacy *see* Greek legacy; Roman legacy
Colford, William 154
compass
and God as architect *9*, 36, 37–40, *47*, *48*, 80, 85, 108, 109, 126
use of 126–7, 128
cord measurement *34*, 91–2, 125, 127
Corpus Agrimensorum Romanorum (Roman surveying manuals) *50*, *51*, *54*, *55*, 86, 110, 111, 116–21, 122, 123, 186
Corpus Christi
and cosmogony, embodiments of 176–84
England, processions and plays *63*, *65*, 134–5, 169–71, 172–3, 177–8, 179–80, 182, 183
Eucharist and cosmic time and space 180–81
Eucharistic processions and routes *60*, *64*, 164–76
Flanders, Bruges, procession of the Holy Blood *60*, 65, *63*, 165–8, 169, 180–81
plays, alleged evolution of 177, 178–9, 181
processions and unification of urban social body 171
and urban embodiment 158–64
Cosmas Indicopleustes, *Christian Topography* 7, 31–2
cosmic and urban space-time, convergence of 176–84
cosmimetry 38, 40, 95, 96, 118, 123
cosmogony
and city imagined 36–40
embodiments of, and Corpus Christi 176–84
cosmogony, lordship and urban foundation 77–85
architects, urban design and geometrical knowledge *9*, *32*, *33*, *48*, 86–90, 109
Christian kingship and sovereign rule 78–9
cord measurement *34*, 91–2, 125, 127
cosmological parallels 94

geometry and urban formation 85–94
lord, role of *31*, 82–5
mensores and town surveys *34*, 90–94
symbolic role of kingship and lordship 79–80, 81, 84
town planning, different levels of 93
town-founding parallel with foundation of world 77–8, 80–81
cosmographies, medieval and practical geometries *see* practical geometries and medieval cosmographies
cosmos-city-body (tripartite scheme) 8–9, 10–11, 12, *46*, 107, 132–3, 135, 137, 138
Creation miniatures 38
cross, significance of 34, 35, 53
Crusader maps 16, 18–20, 21, *41*, 102

Daniel of Morley, *Liber de naturis* 126
Dominicus de Clavasio 115
Geometria Culmensis 92–3
Practica Geometriae 92, 123

east-central Europe
bastides, *place* position 49
Bratislava, circular shape 21
České Budějovice 63
circle and square designs 21, *22*, 59
four-squared layout *19*, 45–6, 51–2, 63
Moravská Třebová 63
Neubrandenburg, circle and square design 59
sovereign lord and founding of new towns 84
Strzelce Krajeńskie, circle and square design *22*, 59
urban laws, models of 146
Ebstorf map 31, 36, *44*, 105, 163–4
Edward I 82, 84, 91, 150
England
Battle, abbey and new town 84
Beverley, Corpus Christi play 135, 170–71
Beverley, municipal self-perception 135, 138, 153–4
Bristol, circle with square form 20–21, *42*, 56–7, 103
Bristol, *Little Red Book* and marginal groups 154–6
Bristol, municipal self-perception 133–5, 138, 139, 143
Bristol, Norman laws, effect of *59*, 149, 151
burhs see burhs of Anglo-Saxon England
Bury St Edmunds, proportional design layout *30*, 68, 90
charters, issuing of 146
Chester, Corpus Christi plays 178, 180, 182
Chester, cross-plan 23–5, 53
Chester, *De Laude Cestrie* (description of city) 23–4, 25, 27, 28, 53, 131–2
Chester, 'mystery plays' 171–2
Chichester, *burh* 43
circle with square form 20–21, *42*, 56–7, 103
Corpus Christi plays *63*, *65*, 135, 170–71, 172–3, 177–8, 179–80, 182, 183
Coventry, Corpus Christi plays 174, 178
Cricklade, *burh* *11*, 43, 62, 63
Exeter, *burh* 42
Exeter, laws regarding newcomers 147
Hereford, and Law of Breteuil *57*, 147–9, 150
Hereford, and marginalization of English burgesses 151
Hereford world map *2*, 16, 36
and Law of Breteuil *56*, *57*, 146–9, 150
lepers and prostitutes, marginalization of 153–7
London, and Carolingian Aachen, similarities between 44
London, *Liber Custumarum* 135–8, 139
London, municipal self-perception 143
Ludlow, orthogonal layout 63
Newcastle, Corpus Christi plays 180
and Norman laws *58*, 146–7, 149–50, 151–2
Norwich, and laws regarding newcomers *58*, 147
Norwich and marginalization 156

Norwich and Norman laws, effect of 149, 151
Nottingham, Danish borough 43
Nottingham, and Norman laws, effect of 149–50, 151
Oxford, *burh* 43, 62
Oxford, municipal seal 21
Shrewsbury, and laws regarding newcomers 146
Shrewsbury, municipal seal *4*, 22
Solent, proportional design layout of towns along 67
Southampton, and marginalization 156
Southampton, and Norman laws, effect of 149
Stamford, *burh 12*, 43
tretis of geometri 124
urban law and newcomers to towns 146–9
Wallingford, *burh* 43, 62
Wareham, *burh 11*, 43, 62
Winchelsea, founding of 82, 84
Winchester, *burh* 42
Worcester, municipal seal *4*, 22
York, Corpus Christi plays *63*, *65*, 172–3, 177–8, 179–80, 183
York, and laws regarding newcomers 147
see also Wales
Eucharist and cosmic time and space 180–81
Eucharistic processions *60*, *63*, 134–5
see also Corpus Christi
Euclid 33, 113
Elements 92, 114–15
Eustache de Beaumarchais 49, 68
exclusionary practice and the body politic 150–52, 153–7
Ezekiel 15, 125–6, 127–8

feast days *see* Corpus Christi
ferramentum, use of 119, 120
Fibonacci, Leonardo, *Practica geometriae 23*, 70–71, 92, 115, 124
Fioraventi, Neri 89
FitzOsbern, William 147, 150, 151
Flanders
Ardres, new town defences, design of 89, 91
Bruges, circular-shaped outer defensive circuit 53
Bruges, measurers 90–91
Bruges, procession of the Holy Blood *60*, *61*, 67, 165–8, 169, 180–81
Fortuna's wheel 36–7
four
cardinal points, 'Adam' letters signifying 132
four-square geometries and their symbolism *19*, 42–53, 63
number, significance of 34, *45*, 106
France
Aigues-Mortes, *bastide* 46
Aix-en-Provence, Le Fête-Dieu processional route *64*, 175
Baa *bastide* 91–2
bastides, formation of *see bastides*, France
Breteuil, Law of *56*, *57*, 146–9, 150
Breteuil-sur-Iton, proportional rectilinear layout *28*, 63
Chartres cathedral school 8, 35
circular built forms *21*, 56
Corbie 116
Créon, *bastide 15*, 57, 63
Dijon, *Historiarum* (city description) 27–8
Fourcès, circular built forms 56
Gascony, sovereign lord and founding of new towns 84
Geaune, *bastide* 51, 57
Grenade-sur-Garonne, *bastide*, proportional layout *17*, 49–50, 63, 68–9, 90, 93
Languedoc, circular built forms 56
Laon cathedral school 35
Le Petit Andely and Lisieux, proportional design layout *30*, 68
Marciac, *bastide 14*, 46
Marseilles, processional route 175
Miramont de Guyenne 67
Mirande, *bastide 18*, 51
Molières, *bastide 14*, 47
Monflanquin, *bastide 22*, 57–9, 67
Monpazier, *bastide 15*, *16*, 46, 63, 67
Montpellier, processional route *64*, 175
Normandy, orthogonal forms 63, 68
Normandy, proportional design layout *30*, 68
Paris, Bibliothèque Nationale manuscript 16

practica geometriae 53
Sarrant, circular built forms *21*, 56
sovereign lord and founding of new towns 84
Strasbourg Cathedral, elevation drawing *32*, 87
Vianne, *bastide 13*, 46
Villeneuve-sur-Lot, orthogonal layout *25*, *26*, 63
Frontinus 120

Genesis, Book of 37, 78, 79, 84
Geometria Culmensis 123
geometrical instruments *see* surveying instruments and cosmological symbolism
geometries
and beauty 62
Florentine 'sine geometry' 70–71
geometric practice and microcosmic symbolism 124–8
geometrical built forms and aesthetics *see* aesthetics and geometrical built forms
and medieval cosmographies, *ars geometriae et arithmeticae* (gromatic texts) 114–15, 117, 118, 119
in non-European religions 186
numerical ratios *30*, 61, 65, 66, 67–9, 72, 90
practica geometriae texts *35*, 53, 92–3, 95–116, 118, 121, 122–3, 124
theoretical and practical 113–16
and urban formation *9*, *32*, *33*, *48*, 85–94, 109
Gerbert d'Aurillac 28, 53, 86, 124
geometriae 114, 116, 117, 121, 122, 123, 124–5
Germany
Aachen 44, 116
Breisach-am-Rhein, orthogonal layout *28*, 66–7
Corpus Christi plays and processions *62*, 169, 177
Frankfurt, Whitsuntide play 182–3
Kenzingen, proportional rectilinear layout 66–7
St Clemens chapel, Schwarzrheindorf, cord measurement *34*, 91
Würzburg, Corpus Christi procession *62*, 169
Gesta Stephani 153
Giovanni di Lapo Ghini 89
gnomon, use of *55*, 120, 121–3
God as architect *9*, 36, 37–40, *47*, *48*, 77, 80–81, 85, 108, 109, 126–7
Gothic cathedrals 68, 72, 73
Greek legacy 28, 115, 159
Adam, letters signifying four cardinal points 132
see also Roman legacy
Gregory of Tours, *Historiarum* (description of Dijon) 27–8
groma, use of 119, 120, 123–4
Grosseteste, Robert 71
Praxis Geometriae 72
Gundissalinus, Dominicus, *De divisione philosophiae* 115, 116, 117

Hebrews, Book of 73, 78, 126–7
Hereford world map *2*, 16, 36
Hildegard of Bingen 22–3, 31
Horn, Andrew 137, 138, 142, 143, 153
Hugh of St Victor 34, 39
Didascalicon 61, 72, 73, 126
Practica geometriae 35, 95–6, 115, 117, 121, 122–3, 124–5, 126
Hyginus 119, 120, 121, 123

Ireland
Armagh, circular monastic town *20*, 54, 55–6
Book of Mulling *20*, 55
Kells, monastic town 54
Synodus Hibernensis 54
Isaiah 24, 72, 126
Isidore of Seville 28, 86, 114
De Rerum Natura 34, 36, *46*, 107
Etymologies 31, 36
Italy
Bobbio 116, 121
circle and square forms *23*, *24*, *49*, 56, 57, 59, 63, 82, 89, 92, 110, 126
Cittadella, circle and square design 59, 81
city-republics 81–2, 138, 142
Florence, straight streets 73
Florence, in tripartite terms 132–3
Florentine 'sine geometry' 70–71
Giglio Fiorentino, proportional design layout 67, 81, 89

mensuratores 90
moral-political writing 139–42
Padua, city description 25–7
Talamone, circle with square form *49*, 56, 57, 59, 89, 110, 126
Talamone harbour 21
Terranuova, circular and square forms, combination of *23*, *24*, 59, 63, 81, 92
Terranuova, proportional design layout 70–71, 93

Jerusalem
circular shape 2, *3*, 15, 16–19, 20, 22–3, 27, 37, *36*, *38*, *41*, *44*, 54, 56, 97, 99, 102, 105
circular and square shapes, reconciling 7, *8*, 18, 20, 21, 25, 32, 35, 56
defence against the Infidel 19–20
heavenly city 15–16, 45, 53, 56
imitation of, in Europe 20–21, 22–3, 25, 26, 27, *42*, 46, 53, 57, 103
square-shaped 16, 18, 20, 25, 26, *37*, *39*, 44, 46, 52–3, 98, 100
as symbolic centre of world 15, 16, 23, 37, 40, 78
via Dolorosa and symbolic significance of ritual performance 167
John of Montpellier (Robertus Anglicus), *Commentary* 113, 124
John of Sacrobosco, *Sphere* 113
Treatise on the Quadrant 113
John of Salisbury, *Policraticus* 79, 142
John of Viterbo, *De Regimine Civitatum* 140, 141–2

Lambert of Ardres 91
Lambert of Saint-Omer, *Liber floridus* 16, *40*, 101
Latini, Brunetto, *Li Livre dou Tresor* 136–7, 138–9, 140, 142, 144, 152
Leviticus 155
limites (lines) as street layout, cosmic inspiration of *55*, 119–20, 121–2, 123
lordship *see* cosmogony, lordship and urban foundation
Lucian, *De Laude Cestrie* (description of Chester) 23–4, 25, 27, 28, 53, 131–2

Macrobius 28, 72, 78, 86
Commentary on the Dream of Scipio 5, 11, 29–30, 96
Mallorca, Jaime II and Petra new towns *19*, 51
mappaemundi 30–31, 36, *36*, *44*, 97, 105, 132, 163–4, 177
Marsilius of Padua, *Defensor pacis* 140–41, 144
mensores and town surveys *34*, 90–94
monastic towns *20*, 54–6
moral topographies 131–57
body metaphors 135, 142–3, 144
and Classical political sources 142
corporate body of citizens, laws pertaining to 141
cosmos and heavenly city 5, 10–11, *36*, *38*, 97, 99, 145
exclusionary practice and the body politic 150–52, 153–7
and hierarchical social differences 144, 150
ruling bodies and divine orders 138–43, 153, 155, 156
social 'deviants', exclusion of 153–6
and urban charters and divine rule 143–4
urban embodiment and the body politic 131–43, 161–2
moral topographies and urban laws
marginalization and placed identities 150–52, 153–5
marginalization and street names 156
models of 146
and moral ordering of 'celestial city' 153
and ordering townspeople 143–57
and ordering townspeople, maintenance of social order and difference 144, 149–52, 153
ordering townspeople, and urban development 144–5, 146
and social distinctions 145–6
urban embodiment and urban governance 133–8, 143
municipal seals *4*, 21–2

Neoplatonism 18, 25, 45, 82, 114–15
architectural metaphors 39, 80–81, 126
Calcidius, *Commentary on Plato's*

Timaeus 6, 8–11, 12, 29, 33, 142
Christian thinking prominence in 8, 9, 81, 114
and founding of cities 78, 79, 84
Macrobius, *Commentary on the Dream of Scipio* 5, 11, 29–30, 96
Plato's *Timaeus* 7–8, 10–11, 28–9, 33, 34, 35, 36, 37, 38, 61, 72, 79, 80, 84
and proportionality 72
Nicomachus 61
number symbolism 34, *45*
numerical ratios *30*, 61, 65, 66, 67–9, 72, 90

orbis quadratus 33, 36, *46*, 107
orthogonal forms *25*, *26*, 62–5, 66–7, 68

Paris, Bibliothèque Nationale manuscript 16
Paris, Matthew, *Chronica Majora* 20
performing bodies 158–84
and celestial symbols, placing of 182–3
city as a 'living' world 163–84
community and communion 159–63
Corpus Christi *see* Corpus Christi
cosmic and urban space-time, convergence of 176–84
cosmological symbolism of performed civic rituals 162–3
Eucharist and cosmic time and space 180–81
geographies of performance 159, 163–76
Rogation Day processions 169
social and spatial parallelism 162–3, 167–8, 169–70, 171–5
symbolic significance of ritual performance 159–62, 164, 165, 166–8, 174, 177–80, 181–4
Philo of Alexandria 38–9, 77, 78
planets, trajectory of *43*, 104
planimetry 40, 95, 96, 118, 123, 124
Plato 186
Timaeus 7–8, 10–11, 28–9, 33, 34, 35, 36, 37, 38, 61, 72, 79, 80, 84
practica geometriae texts *35*, 53, 92–3, 95–116, 118, 121, 122–3, 124
practical geometries and medieval cosmographies 95–119
ars geometriae et arithmeticae (gromatic texts) 114–15, 117, 118, 119
Corpus Agrimensorum Romanorum (Roman surveying manuals) *50*, *51*, *54*, *55*, 86, 110, 111, 116–18, 119–21, 122, 123, 186
and *practica geometriae* texts *35*, 53, 92–3, 95–116, 118, 121, 122–3, 124
scholastic learning and the cosmos 114, 115, 116, 118
surveying instruments *see* surveying instruments and cosmological symbolism
theoretical and practical geometries 113–16
Pratike de Geometrie 96–113
proportioned forms *15*, *17*, *22*, *27*, *28*, *29*, 49–50, 61–2, 63, 65–72, 82, 89, 90, 93
root 2 proportions *30*, 68–9, 90
Proverbs 126
Psalter map 15, 16, 31, *36*, 97
Ptolemy of Lucca 61, 81–2
De Regimine 140, 141
Pythagoras 61, 72, 81, 113

quadrant, use of *52*, 112, 124, 125
quadrivium 25, 61, 113, 114, 115, 116

Rabanus Maurus 32–3
reed (rod), use of 127–8
Remigio de' Girolami 132–3
Revelation, Book of 15, 25, 26, 27, 35, 36, 37, 53, 78, 127, 128
Ricart, Robert 20–21
'Maire of Bristowe is Kalendar' *42*, 103, 133–5, 138, 143, 154, 155
Richard of St Victor 39, 125, 127
Robertus Anglicus (John of Montpellier), *Commentary* 113, 124
Rogation Day processions 169
Roman legacy
castra 44
colonia *18*, *19*, *50*, 110, 121
Corpus Agrimensorum Romanorum (Roman surveying manuals) *50*, *51*, *54*, *55*, 86, 110, 111, 116–21, 122, 123, 186
see also Greek legacy

scholastic learning and the cosmos 114, 115, 116, 118
Sibylline Oracles 132
social distinctions, and urban laws 145–6
social order and difference, maintenance of 144, 149–52, 153
Socrates 11
Spain, orthoganal urban layouts 63
square built forms
burhs of Anglo-Saxon England *10*, *11*, 42–5, 51, 52
in Christian imaginings of the world 31–4
circle and square, combining, urban landscapes and symbolic forms *21*, *22*, *23*, *24*, 56–60
England, circle with square form 20–21, *42*, 56–7, 103
four-square geometries and their symbolism *19*, 42–53, 63
groma, use of (square-based plans) 119, 120, 123–4
Italy, circle with square form *23*, *24*, *49*, 56, 57, 59, 63, 82, 92, 110, 126
Jerusalem 16, 18, 20, 25, 26, *37*, *39*, 44, 46, 52–3, 98, 100
Jerusalem, circular and square shapes, reconciling 7, *8*, 18, 20, 21, 25, 32, 35, 56
rotating square method 68, 69–70
see also circular built forms
streets
limites (lines) as street layout, cosmic inspiration of *55*, 119–20, 121–2, 123
names and marginalization 156
straighter *15*, *25*, 63, 72–3
surveying instruments and cosmological symbolism 119–28
altimetry 40, 95, 96, 118, 123, 124
astrolabe, use of 40, *53*, 112, 124–5
Bible references 125–7
compass, and God as architect *9*, 36, 37–40, *47*, *48*, 80, 85, 108, 109, 126
compass, use of 126–7, 128
ferramentum 119, 120
geometric practice and microcosmic symbolism 124–8
gnomon, use of *55*, 120, 121–3
groma, use of 119, 120, 123–4
limites (lines) as street layout, cosmic inspiration of *55*, 119–20, 121–2, 123
linking heaven and earth 120–24
planimetry 40, 95, 96, 118, 123, 124
quadrant, use of *52*, 112, 124, 125
reed (rod), use of 127–8
and Roman *colonia* *18*, *19*, *50*, 110, 121
tretis of geometri 124
Switzerland
Berne, proportional rectilinear layout *27*, *29*, 66–7
Lucerne, Passion Play 183
St Gall monastery 121
symbolic roles
aesthetics and geometrical built forms 64–5, 68, 72–3
circular geometries and *20*, 53–6
cosmological symbolism of performed civic rituals 162–3
four-square geometries and their symbolism 42–53
geometric practice and microcosmic symbolism 124–8
Jerusalem as symbolic centre of world 15, 16, 23, 37, 40, 78
kingship and lordship 79–80, 81, 84
number symbolism 34, *45*
performing bodies and celestial symbols, placing of 182–3
significance of ritual performance 159–62, 164, 165, 166–8, 174, 177–80, 181–4
surveying instruments and cosmological symbolism *see* surveying instruments and cosmological symbolism
urban landscapes and symbolic forms *see* urban landscapes and symbolic forms
via Dolorosa and symbolic significance of ritual performance 167

Templum Domini *2*, *3*, 19, *41*, 102
Thierry of Chartres 38
Trier manuscript 16, 18, *37*, 98
Trinity Apocalypse manuscript 16, *39*, 100
tripartite scheme (cosmos-city-body) 8–9, 10–11, 12, *46*, 107, 132–3, 135, 137, 138

urban
charters and divine rule 143–4
and cosmic space-time, convergence of 176–84
design and geometry *9*, *32*, *33*, *48*, 85–94, 109
development, and ordering townspeople 144–5, 146
embodiment and the body politic 131–43, 161–2
embodiment and urban governance 133–8, 143
foundation, lordship and cosmogony *see* cosmogony, lordship and urban foundation
laws and moral topographies *see* moral topographies and urban laws
town planning, different levels of 93
town-founding, parallel with foundation of world 77–8, 80–81
urban landscapes and symbolic forms 41–60
bastides see bastides, France
burhs see burhs of Anglo-Saxon England
Christianity and other religions, shared beliefs 186–7
circle and square, combining *21*, *22*, *23*, *24*, 56–60
circular geometries and their symbolism *20*, 53–6
four-square geometries and their symbolism 42–53
monastic towns *20*, 54–6
urban mappings
Christian imaginings of medieval world 28–35
city description 23–8
cosmogony and city imagined 36–40
heavenly to earthly city 15–28
holy city depiction 18–23

Valenciennes manuscripts 16, 17, 18, *38*, 99
Vernani of Rimini, Guido, *Commentary on the Politics* 140
Villard de Honnecourt, *Sketchbook 33*, 68, 87–8, 125

Wales
colonization and exclusion of residents 150
Flint and Rhuddlan, founding of 82–3
see also England
Wichmann of Magdeburg, Archbishop 84
William of Conches 10–11, 29, 38, 142
Commentary on the Timaeus 126

Zähringer dynasty, proportional rectilinear layout of 'new towns' *27*, *28*, *29*, 66–7